MW00837080

Visible and Dark Matter in the Universe

This is a concise introduction to modern astrophysics for physicists, with a focus on galaxy dynamics and the discovery of dark matter halos in galaxies. Part I summarizes important discoveries in observational astronomy and astrophysics, in a manner accessible to those who are new to the topic. Building on this foundation, Part II describes the study of dark matter and provides more detail on galactic dynamics. Important physical concepts that form the basis of key astrophysical phenomena are explained, avoiding unnecessary technicalities and complex derivations. The approach is semi-empirical and emphasizes the importance of key measurements and observations in formulating fundamental theoretical questions and developing their solutions. Students are encouraged to develop a deep understanding of major discoveries and contemporary research topics, beyond the simple application of practical models and formulae, as a bridge to more advanced study in astrophysics.

GIUSEPPE BERTIN is Professor of Physics at the University of Milan, Italy. He has previously held several positions at the Scuola Normale Superiore and the Massachusetts Institute of Technology, and has been a member of the Kavli Institute for Theoretical Physics, University of California, Santa Barbara. His previous books include *Spiral Structure in Galaxies: A Density Wave Theory* with C. C. Lin (MIT Press) and *Dynamics of Galaxies* (Cambridge University Press). He was elected to Italy's Accademia Nazionale dei Lincei in 2013, when he received its Premio Nazionale del Presidente della Repubblica in Science.

Visible and Dark Matter in the Universe

A Short Primer on Astrophysical Dynamics

GIUSEPPE BERTIN

University of Milan

Shaftesbury Road, Cambridge CB2 8EA, United Kingdom

One Liberty Plaza, 20th Floor, New York, NY 10006, USA

477 Williamstown Road, Port Melbourne, VIC 3207, Australia

314–321, 3rd Floor, Plot 3, Splendor Forum, Jasola District Centre, New Delhi – 110025, India

103 Penang Road, #05–06/07, Visioncrest Commercial, Singapore 238467

Cambridge University Press is part of Cambridge University Press & Assessment, a department of the University of Cambridge.

We share the University's mission to contribute to society through the pursuit of education, learning and research at the highest international levels of excellence.

www.cambridge.org
Information on this title: www.cambridge.org/9781316519318

DOI: 10.1017/9781009023368

© Giuseppe Bertin 2023

This publication is in copyright. Subject to statutory exception and to the provisions of relevant collective licensing agreements, no reproduction of any part may take place without the written permission of Cambridge University Press & Assessment.

First published 2023

A catalogue record for this publication is available from the British Library.

A Cataloging-in-Publication data record for this book is available from the Library of Congress

ISBN 978-1-316-51931-8 Hardback

Additional resources for this publication at www.cambridge.org/bertin

Cambridge University Press & Assessment has no responsibility for the persistence or accuracy of URLs for external or third-party internet websites referred to in this publication and does not guarantee that any content on such websites is, or will remain, accurate or appropriate.

To my brother Alberto, a physics teacher with great
interest in science and astronomy

Contents

Preface

The structure of this book reflects the first part of a one-semester course addressed to third-year undergraduate students of physics at the University of Milan. In the form taken in the last 15 years, the course contained two additional parts. The second part was taught by a colleague active in the study of the cosmic microwave background radiation, who introduced several aspects of observational astronomy and observational cosmology (recently this contribution was replaced by two colleagues briefly covering the topics of black holes in astrophysics and modern aspects of cosmology). The third part was given by a colleague working on neutrino experiments at Gran Sasso, who focused on topics in astroparticle physics.

My task was to set the general themes for such a composite set of lectures, which were meant to help undergraduate students, with no previous experience in astronomy courses, to become familiar with key research topics in astrophysics and possibly to decide, on the basis of these lectures, whether they would wish to undertake a program in astrophysics for their future graduate studies.

Within this general framework, in my part (20 hours of class) I tried to pose and answer the following key questions:

(i) What are the most significant discoveries in astrophysics of the last 100 years?
(ii) What does research in astrophysics consist in?
(iii) Why does gravity, even at the classical level, still pose some of the most intriguing and challenging problems in modern astrophysics?

To proceed, and to do so at an elementary level within a very limited number of classes, I decided to focus on few selected topics that I judged to be particularly significant and instructive. I divided my time-slot in two equal parts. The first was devoted to providing highlights of major discoveries, as an answer to the first key question mentioned above. Then, in order to be more specific and to tackle at least one topic in some depth, in the second part of my slot I described the discovery of dark matter from the study of galaxies and clusters of galaxies. This allowed me to provide an answer to the third key question posed above. I thus produced a personal view of introducing astrophysics to students of physics. My classes were all based on concrete examples, in which I tried to convey my firm conviction that progress and inspiration come from few decisive measurements/observations rather than from abstract or deductive arguments.

The interactions with the students revealed a number of apparently surprising characteristics of our physics program. While generally showing a rather satisfactory background in modern physics (in particular, many already showed a reasonable familiarity with basic concepts in quantum mechanics and general relativity), the students often, almost always, exhibited a rather weak and naive attitude toward classical mechanics and related tools of calculus. Given the fact that my research activity focuses on the dynamics of galaxies and other stellar systems, I thus found it natural to take the task of presenting a short introduction to astrophysics as an excuse to insert, here and there, essentially in each class I gave, some simple problems or a brief outline of some basic concepts in classical mechanics. (In general, this supplementary material corresponds to the final section of each chapter.) In doing so, I tried to prompt the students to catch up with topics that not too long ago would be taken for granted as first-year physics. These digressions in classical mechanics not only turned out to be useful, but also helped excite some interest in dynamical aspects of astrophysics. In other words, to some students they may have shown that a view of astrophysics from the dynamical window has its own beauty.

For obvious reasons, the 10 chapters of this book expand significantly what I could teach realistically in 10 two-hour classes, and thus this text may be taken as the basis for a full one-term course.

I should also mention that some of the topics covered in this short book draw from material to some extent present in two books that I wrote. However, there is a clear contrast in emphasis and content, which here are fully aimed at communicating with undergraduate students that are imagined to meet astronomy and astrophysics for the first time. In turn, the book *Dynamics of Galaxies* (2nd ed., Cambridge University Press, 2014) is addressed to graduate students, with the goal of providing a coherent framework for our current perception and tools of investigation on the subject of the dynamics of galaxies, while the monograph *Spiral Structure in Galaxies: A Density Wave Theory* (co-authored with C. C. Lin; MIT Press, 1996) has a specific objective, that is, to provide a coherent framework for understanding spiral structure in galaxies.

Of course, in class I make wide use of the blackboard and plots taken from books and other published material. In the current text, there are only a few figures, most of them taken from seminal publications. My hope is that this apparent limitation could serve as an incentive for the student to look up the original references and other sources so as to properly complement the text with the material that is generally provided in class. I am especially keen on encouraging students to look up original references, which are selected and listed in the relevant endnotes.

Finally, I would like to mention another point that is complementary to the role of my teaching in writing this book. In Milan, at the end of their third year, undergraduate students write an essay (in Italian, *Tesi di Laurea Triennale*), for which the student has no obligation of obtaining original results as would be suitable for a scientific publication or for a Ph.D. thesis. For the students, the essay is a way to get exposed, often for the first time, to scientific research; that is, to get an answer to the second key question that I raised above. So far I have supervised about 20 such projects on topics of astrophysical interest. The interactions with students during the supervision process have had a significant influence in determining the content and the structure of the present book.

Acknowledgments

It would be impossible to thank properly all the scientists who have contributed to the ideas and the results presented here. The Department of Physics, University of Pisa, is thanked for their hospitality, which has given me the opportunity to write part of this book.

PART I

Visible Matter

1
Light

Electromagnetic radiation is the primary source of astronomical information. In particular, until the early 1930s all astronomy was based on the use of telescopes that extended the power of the human eye but were restricted to the collection of visible light. Then the advent of radio astronomy marked the beginning of a revolution, which later bloomed when the space age, starting in the late 1950s, made it possible to observe the sky from devices operating outside the atmosphere of our planet. This and the development of new technological tools soon allowed us to exploit wider and wider intervals of the entire spectrum of the electromagnetic radiation as a way to probe the properties of the universe. In general, the sources of astronomical electromagnetic radiation and other sources of astronomical information (see Chapter 5) are what we call visible matter.

The purpose of this chapter is to introduce some key concepts and notation that characterize light and the collection of light for astronomical purposes. We will also briefly outline some obvious complications that affect the acquisition of observational data, some of which are intrinsic to electromagnetic radiation, others to the telescopes and instruments that are used, and, for observations from the ground, the complications related to the presence of the atmosphere. We will then proceed to a brief description of the main types of information that we may extract from the observations, by means of imaging and spectroscopy. We will recall the difference between apparent and intrinsic properties of the astronomical sources, which is at the basis of probably the most important problem in astronomy, that is, the measurement of the distance to a given source. We will then comment on the fact that the light from distant sources is often a mixture of photons from different stars or different components. This will serve as an excuse for a quick introduction to important concepts, such as stellar populations, mass-to-light ratios, mean motions, and velocity dispersions.

In closing the chapter, we will describe a method to measure the distance to a stellar system based on the application of a very simple dynamical model to a suitable set of observations.

1.1 The Electromagnetic Spectrum, Imaging, and Spectroscopy

1.1.1 Types of Radiation and Wavebands

The electromagnetic spectrum is divided into broad regions (defined in terms of photon energy E or, equivalently, of photon wavelength λ or frequency ν). We recall that from the relation $E = h\nu = hc/\lambda$, where $h = 6.6261 \times 10^{-27}$ erg s is the Planck constant and $c = 2.9979 \times 10^{10}$ cm s^{-1} is the speed of light in vacuum, the wavelength associated with 1 eV $= 1.6022 \times 10^{-12}$ erg is 1.2398×10^{-4} cm $= 1.2398\ \mu = 1.2398 \times 10^{4}$ Å $= 1.2398 \times 10^{3}$ nm. This also sets the relation between the often used units micron (μ), angstrom (Å), and nanometer (nm).

The gamma ray domain refers to photon energies greater than 1 MeV and wavelengths of 1 Å or smaller. X-rays have lower energies, down to energies $\approx$ 1 keV (i.e., wavelengths from 1 to 100 Å); soft X-rays are those with lower energies, below $\approx$ 10 keV, and hard X-rays have higher energies. The ultraviolet (UV) part of the spectrum extends from wavelengths in the range 100 Å to $\approx$ 4000 Å. Visible light covers the wavelength interval of 4000–7000 Å. Then infrared radiation is characterized by wavelengths below $100\ \mu = 10^{-1}$ mm; in particular, near-infrared photons have wavelengths of one or few microns, whereas at longer wavelengths astronomers talk about far-infrared radiation. Finally, radio waves are those with wavelengths of millimeters or larger (in particular, those with wavelength up to 1 m are often called microwaves).

Observations in one of the above-defined broad regions of the electromagnetic spectrum are often subdivided into finer regions, called wavebands. In particular, visible light, which is the focus of all observations before the advent of modern astronomy, is often divided into bands, such as B, V, R, I, whereas in more modern near-infrared observations we distinguish between J, H, K bands in the order of increasing wavelength. These subdivisions often reflect commonly used filters in astronomical observations and may correspond to specific transparency windows in the atmospheric transmission.

1.1.2 Atmospheric Transparency

The atmosphere is basically transparent to visible light and to radio waves with wavelengths larger than 1 cm up to 10 m. It is basically opaque to high-energy

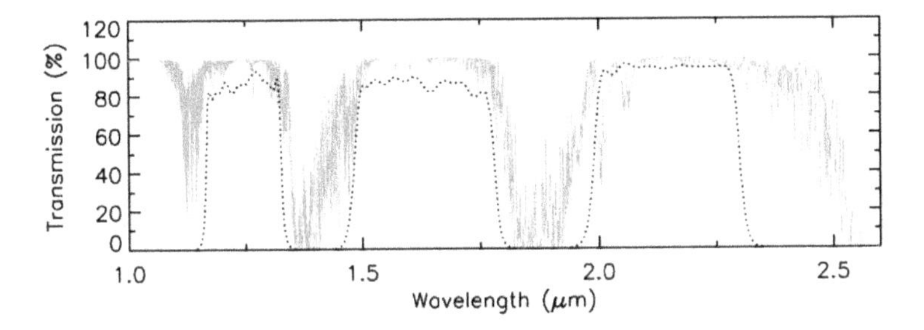

Figure 1.1 From left to right, J, H, and K filter profiles (dotted lines) superposed on the atmospheric transmission at Mauna Kea (From: Tokunaga, A. T., Simons, D. A., Vacca, W. D., "The Mauna Kea Observatories near-infrared filter set: II. Specifications for a new JHKL'M' filter set for infrared astronomy," 2002. *Publ. Astron. Soc. Pacific*, **114**, 180; © The Astronomical Society of the Pacific. Reproduced by permission of IOP Publishing. All rights reserved.).

incoming radiation, from the UV to gamma rays. In the astrophysically interesting domain of the near-infrared and millimetric radiation, there are several excellent transmission windows, which are best exploited by telescopes located at high altitudes. This is one important reason that explains why many observatories have been built at relatively high altitudes.

The atmospheric transparency is often illustrated quantitatively by plotting, as a function of the radiation wavelength, the altitude at which the intensity of the radiation coming from an astronomical source is reduced by a factor of 2. Alternatively, at a given astronomical site, we may plot the transmission as a function of wavelength, with the standard definition that the transmission is taken to be unity if the intensity of the incoming radiation is unaffected (see Fig. 1.1).

1.1.3 Hydrogen Lines

The fact that most of the visible matter in the universe is made of hydrogen suggests that a large fraction of what we can extract from astronomical observations derives from the identification of hydrogen lines, produced either in emission or in absorption by transitions involving different energy levels of the hydrogen atom. In particular, it is well known that the lines of the Lyman series, that is, of transitions from various excited levels to the ground level, fall in the UV part of the spectrum and that the Balmer series, associated with the transition from higher levels to the first excited level, fall in the visible, whereas the Paschen and Brackett series fall in the infrared. In the ultraviolet, a special role is played by the Lyα line connecting the first excited level to the ground level, with wavelength 1216 Å. In the visible, part of the Balmer series, important lines are Hα at 6563 Å and Hβ at 4861 Å.

Of course, the above information refers to the source. Because of the cosmological redshift, the Lyα emission of distant sources can be brought into the visible part of the spectrum of the observer, whereas the Balmer lines may be moved to the near-infrared.

In Chapter 3, we will see that a major development occurred in radio astronomy when, first theoretically, and then observationally, it was discovered that a hyperfine transition related to the spin alignment of the electron and the proton in atomic hydrogen is associated with a line at 21.106 cm (1420.4 MHz), often referred to as the 21-cm line.

1.1.4 Telescopes

Larger and larger telescopes have been built and are being planned, with the goal of studying the sky in better and better detail and of gathering information on fainter and fainter sources. This technological progress is accompanied by major developments in the creation of light and optimally performing materials for the construction of the telescope parts, in the construction of instruments, and in data storage and analysis. For a long time the largest optical telescopes were the 5-m (200-inch) Hale telescope, located at Palomar mountain in California and operative since 1949, and the 6-m (20-ft) telescope at the Special Astrophysical Observatory, in the Russian Caucasus mountains, operative since 1975. The specifications 5 m and 6 m denote the diameter of the so-called primary mirror; they are often referred to as aperture of the telescope. It is commonly perceived that large telescopes are built because larger telescopes have better angular resolution. However, this is only partly true.

1.1.5 Angular Resolution and Sensitivity

The angular resolution of an imaging device can be defined as the minimum angular distance between two point sources that can be effectively separated or distinguished by an observation. It can be shown that an ideal system, characterized by an effective aperture D, dealing with electromagnetic radiation of wavelength λ, has an angular resolution $\theta \approx \lambda/D$, which is often called the diffraction limit. Real devices have poorer angular resolution.

The human eye has an angular resolution of $\approx$ 1 arcmin. The star Mizar in the constellation Ursa Major has a closeby star, Alcor, which can be easily resolved with the naked eye (it is at an angular distance of $\approx$ 12 arcmin from Mizar).

As another example of angular scales, we may mention the Galilean satellites, the four brightest moons of planet Jupiter, that are typically located at an angular distance of 2 to 10 arcmin from Jupiter and easily visible with normal binoculars. Io's angular diameter is ≈ 1 arcsec.

Sensitivity describes the measure of the faintest signal that can be detected. For a telescope it scales as the square of its aperture D^2, that is, of its collecting area in relation to the incoming photons. Of course, another parameter that is involved in determining the sensitivity of an observation is the exposure time. With a given telescope deeper images, that is, images that reveal fainter details, are generally obtained by taking longer exposures.

In terms of sensitivity, under very good conditions the human eye is able to see stars ≈ 100 times fainter than the brightest star of the constellation Leo, the blue star Regulus.

1.1.6 Seeing and Point Spread Function

A primary factor that severely limits optical observations from the ground is the turbulence present in the atmosphere, that is, in the air through which incoming photons pass before reaching the telescope. The main properties of the atmospheric turbulence in relation to astronomical observations are the relevant cell size of the air clumps (at visible wavelengths, ≈ 10 to 20 cm) and the typical time scale over which the cell optical properties change ($\approx 10^{-2}$ s or below). Astronomers generally describe the phenomenon by saying that observations from the ground are affected by seeing. Broadly speaking, seeing is the angular diameter (full width at half maximum) of a disk into which a point source is imaged in a relatively long exposure as a result of the blurring effect of atmospheric turbulence. It has been realized that much of the effect is due to the state of the air in the vicinity of the telescope; because of this, modern telescopes are built with special care, especially in relation to the thermal properties of the hosting domes. In practice, a good astronomical seeing is of the order of 1 arcsec. Under exceptional conditions, the seeing at the best observatories can be as low as 0.3 arcsec. For large optical telescopes, this is generally much worse than the diffraction limit. Therefore, the construction of large optical telescopes is mainly justified by their better sensitivity, and only to a lesser extent by their resolving power. However, we will see in the next subsections that astronomers have found ways to bypass, in large part, this limitation to observations.

A simple phenomenon, experienced by the human eye, that is related to the blurring effect of turbulence on visible light is the twinkling of the stars (which are effectively point sources), as opposed to the steady light from the brightest planets (which are angularly small, but finite-size extended sources).

Note that the apparent diameters of planets Venus, Mars, Jupiter, and Saturn are, $\approx$ 10 to 65, $\approx$ 4 to 25, $\approx$ 30 to 50, and $\approx$ 15 to 20 arcsec, respectively. We recall that the Moon's angular diameter is $\approx$ 30 to 35 arcmin.

Even in the absence of effects of turbulence in the atmosphere, a given optical device, because of technical limitations and other factors, reduces the image of a point source to a distorted finite-size spot. Astronomers quantify this general effect by defining the Point Spread Function (PSF), which describes how a point is blurred in the observation. The observed image of an extended source is the convolution of the source signal with such PSF.

1.1.7 Active and Adaptive Optics

Modern telescopes are capable of effectively controlling the surface collecting the light from astronomical sources, so as to overcome some of the effects that would spoil the results of the observations with respect to those that could be obtained under ideal conditions. The process is performed by means of a combination of mechanical tools (actuators in the case of active optics) and electronic tools.

Active optics (starting in the 1980s) generally refers to actions taking place on the time scale of seconds, to compensate for relatively large amplitude mechanical and thermal stresses that may be induced by the geometric configuration of the telescope with respect to gravity and winds.

Adaptive optics (starting in the 1990s) refers to smaller-amplitude actions taking place on the time scale of 10^{-2} s and below, aimed at overcoming the effects of seeing. In its simplest form, the general strategy is to take advantage of a sufficiently bright point source in the field of view (a guide star), during observations, and thus to read off from its distortion the relevant PSF that can be applied to reconstruct the desired seeing-free image. In the most recent versions of adaptive optics, when a sufficiently bright point source is not available in the desired direction, an artificial guide star in the field of view can be created by shining a suitable laser beam toward the sky.

1.1.8 Interferometry

Another way to improve angular resolution is to make use of interferometric techniques. The general idea is to acquire the signal from incoming electromagnetic waves with a set of separate telescopes placed at different locations and to coherently superpose the signals in such a way that the separate telescopes behave as parts of a single detecting device. The largest distance between two telescopes of an interferometric configuration is the largest baseline; if we call it D, the ideal angular resolution achievable by the interferometer is $\approx \lambda/D$. Of course, even if angular resolution can be greatly improved by interferometry, the sensitivity remains limited by the total collecting area of the array of telescopes, which is generally much smaller than D^2.

The technique was developed successfully very soon at long wavelengths, in radio astronomy, starting in the 1940s. It is routinely used with continental and intercontinental baselines of thousands of kilometers (VLBI).

At shorter wavelengths, in the infrared and in the visible, major technological advances are required, and interferometry was developed mainly at the end of the last century. As a notable example, we should mention the Very Large Telescope Interferometer (VLTI), in which the large 8.2-m telescopes of VLT on Cerro Paranal in Chile can be used in interferometric mode together with smaller and mobile 1.8-m auxiliary telescopes, achieving ≈ 200 m as the largest baseline.

1.1.9 Imaging and Spectroscopy

There are two main modes of astronomical observation, imaging and spectroscopy.

Typically, images are intensity maps, in a given waveband, that provide us with morphological details of the selected field of view. For certain sources, such as nearby globular clusters, images give a picture of the way stars crowd up in the central regions. For other, more distant, extended sources, such as galaxies, images give us information about the overall shapes that characterize the sources. Clearly, images are two-dimensional maps (they cover a small solid angle in the sky) for which the intrinsic physical size (length scales) can be set only if we can measure the distance to the source. The problem of inferring the three-dimensional structure of an observed source or field is one of the key open problems of astronomical observations; a related open problem, for an extended source characterized by some internal symmetry, is to determine its inclination with respect to the line of sight.

We tend to interpret the observed morphology in terms of internal structure of the source. However, if we define structure as mass distribution, it is obvious that images in certain wavebands may give us misleading information about the source structure, either because of absorption of the photons in their path from source to detector or, more simply, because the waveband that we are using does not correspond to the source component that best traces the mass distribution. In this respect, for the purpose of studying the structure of galaxies, it has been realized that the best representative images are those obtained in the near-infrared, because this type of radiation is least affected by extinction by interstellar dust and because this emission best traces the evolved red giant stars that are thought to make the bulk of the mass of the stellar component of galaxies; in contrast, images in the visible exhibit morphological details that are best representative of the interstellar medium and newly born stars. In a completely different context, in the visible the intergalactic space in clusters of galaxies appears to be practically empty, whereas we now know from X-ray observations (see Chapter 4) that it contains very large amounts of matter in the form of a diffuse plasma.

Spectroscopy studies the incoming flux distribution as a function of wavelength and is especially valuable when certain lines, either in absorption or in emission, can be identified and analyzed. Spectroscopy is the key tool to make us understand the nature of the mechanism of electromagnetic emission operating in a given source. It provides important information on the physical state of the source (e.g., on its temperature) and on its chemical composition.

One important role of spectroscopic observations is their diagnostic power in relation to kinematics. For those sources for which spectral lines can be identified and analyzed, the Doppler effect provides us with the opportunity to extract information on the velocity of the source relative to the observer along the line of sight. For an observed nearby stellar system at known distance, such as a globular cluster, images give information on two spatial coordinates (in the plane of the sky), and spectroscopy gives information on the velocity component of stars along the line of sight; in other words, the combined use of imaging and spectroscopy gives us information on three out of six of the coordinates that define the relevant phase space.

1.2 Apparent and Intrinsic Quantities, Standard Rods, and Standard Candles

If we do not know the distance to a given source, observations give us only incomplete information about the system that we are studying. In practice, we can measure its apparent size (e.g., if we are observing an extended source, its apparent diameter) and its apparent luminosity (in a given waveband, from the flux that we receive with our telescopes). A given observed source can be rather small and faint, near to us, or, alternatively, it can be huge and powerful, if it turns out to be very far away.

This general theme has set long-lasting landmark controversies about the nature of certain sources that were eventually resolved by a convincing distance determination. Notable examples are the nature of the nebulae and the discovery of galaxies and, in more modern times, the nature of Gamma Ray Bursts and the discovery that they originate in systems located at cosmological distances. The high-energy phenomenon of Gamma Ray Bursts, as will be briefly described in Chapter 4, demonstrates another important aspect that connects apparent and intrinsic luminosity, that is, whether the source is emitting in a beamed way or isotropically over the whole sky.

1.2.1 Magnitudes and Parallaxes

In general, astronomers measure apparent and absolute luminosities in magnitudes and lengths in parsecs or kiloparsecs.

Without facing the task of providing complete and exact definitions, we would only like to mention here that, with respect to the standard units used in physics to measure luminosities, magnitudes correspond to taking the operation $-2.5\log$, where the logarithm is meant to be to base 10. Therefore, if we compare two sources, the first of which is 100 times brighter than the second, their magnitudes differ by 5, and the fainter source has larger magnitude.

The brightest stars in the sky have apparent magnitudes around zero. The brightest, Sirius, shines at $m_V \approx -1.47$ mag in the V band, and, being located at a distance of ≈ 2.64 pc from us, it is characterized by absolute magnitude $M_V \approx 1.42$ mag; that is, this would be its luminosity in magnitudes if Sirius were located at a distance of 10 pc. For comparison, the Sun's apparent magnitude is ≈ -26.7, whereas its absolute magnitude is ≈ 4.83; in more standard units the Sun's absolute luminosity is $1\ L_\odot \approx 3.83 \times 10^{33}$ erg s^{-1}. At its brightest, Venus shines at ≈ -4.9 mag.

The unit of length that is used most frequently is the parsec, 1 pc $\approx 3.09 \times 10^{18}$ cm ≈ 3.26 light-years. The origin of the parsec, and its precise definition, is traced to a process of triangulation, in which the parsec is defined as the distance at which 1 AU $\approx 1.5 \times 10^{13}$ cm (the Astronomical Unit is the distance between the Earth and the Sun) subtends an angle of 1 arcsec. The very small change of position in the sky of a nearby star, in a frame of reference given by much more distant stars, when the observation is made at different times during the Earth's orbit around the Sun, is called parallax. Distance measurements are also sometimes called parallaxes, even when the distance measurement does not involve triangulation.

In closing this subsection, we record some dimensional relations that turn out to be useful in the course of many astronomical calculations. For angles, note that 1 radian is $180^\circ/\pi \approx 57.2958^\circ \approx 206265''$. For times, 1 yr $\approx \pi \times 10^7$ s. For velocities, 1 km s$^{-1} \times 10^6$ yr ≈ 1.02 pc.

1.2.2 Line-of-Sight Velocities and Proper Motions

The Doppler effect is at the basis of velocity measurements in astronomy. Typically, the measurement consists in identifying a certain line (or certain lines) in the light coming from a given source. For relatively small speeds, the line-of-sight velocity v_{los} between the source and the observer is obtained by measuring the wavelength displacement $\Delta\lambda$, that is, the difference between the observed wavelength and the wavelength λ at emission, and by applying the relation $\Delta\lambda/\lambda \approx v_{los}/c$, where c is the speed of light. If the source is receding from the observer, the observed wavelength is longer, that is, it is redshifted. This gives a direct measurement of v_{los}, which is obviously distance independent.

For stars belonging to our Galaxy or nearby galaxies, measurements of this kind give velocities of up to few hundreds of kilometers per second; for stars in the solar neighborhood, that is, within a few hundred parsecs from us, the relative velocity v_{los} of individual stars is often of the order of 30 km s^{-1}. In the nearby universe, that is, for galaxies a few hundred megaparsecs away, the Doppler shift $z = \Delta\lambda/\lambda$ is always a redshift and is (approximately) directly proportional to the distance d of the source, corresponding to the Hubble law $v = H_0 d$. The quantity $H_0 \approx 70$ km s^{-1} Mpc^{-1} is called the Hubble constant and measures the expansion rate of the universe in our cosmological vicinity. The quantity H_0^{-1} is of the order of 10^{10} yr and thus is the basic time scale that sets the age of the universe (the time that has passed since the Big Bang took place). Astronomers now observe galaxies characterized by redshift greater than unity; of course, a simple interpretation of these data in terms of recession speeds along the line of sight is less significant, and a cosmological interpretation well beyond the simple concept of the Doppler shift is required.

Going back to the motion of nearby stars, it is clear that, depending on their distance and on the value of the velocity component transverse to the line of sight, by taking observations at significantly different epochs, several years apart from one another, we may be able to detect the motion of a star in the sky, with respect to a frame of reference provided by much more distant stars. The associated velocity that can be extracted from a set of measurements of this type is called proper motion and is typically given as an angular velocity vector ω, which is only an apparent quantity. Recall that the position of a star in the sky is given by two angles [e.g., astronomers often refer to equatorial coordinates defined by the pair of angular coordinates (α,δ), called right ascension and declination; note that, because of the geometry of spherical coordinates, the angular velocity is related to the time derivatives of the two angles by the relation $\omega = (\dot{\alpha}\cos\delta, \dot{\delta})$]. If we know the distance d to the star of which we have measured the proper motion, we can thus reconstruct the intrinsic transverse velocity $v_\perp = \omega d$ relative to us. In practice, the angular displacements are always

Figure 1.2 The globular cluster NGC 5139 (ω Cen) in an image taken with a field of view of approximately 30 × 30 arcmin from ESO's La Silla Observatory. The image shows only the central part of the cluster. North is up, east is to the left; it is a composite of B, V, and I filtered images (credit: European Southern Observatory "CC BY 4.0."). Currently, globular clusters are believed to contain only small amounts of dark matter, although they are likely to host significant amounts of dark remnants (see Section 3.3). A color version of this figure is available at www.cambridge.org/bertin

very small. Measurements of proper motions generally require long periods of observations with telescopes characterized by the capability of measuring positions with extreme accuracy (astrometric precision). Yet, the development of new telescopes and instruments (e.g., see Chapter 2) and patient analysis combined with a nontrivial discussion of the problem of the relevant frames of reference has already made it possible to measure proper motions and intrinsic transverse velocities for thousands of stars in globular clusters such as ω Cen (see Fig. 1.2).

1.2.3 Standard Rods, Standard Candles, and Distance Measurements

As we mentioned earlier in this section, the simplest distance measurement for stars can be made by triangulation, by taking advantage of the motion of the

Earth around the Sun. But, until recently, this measurement has been possible only for the nearest stars (but see Section 2.3). When triangulation is not feasible, a distance measurement is typically based on the identification of a suitable "standard rod" or "standard candle." With these terms we indicate astronomical objects of which we have independent convincing evidence for either their intrinsic linear size or their intrinsic luminosity; in modern astronomy, an example of a standard candle is a special type of supernova (type Ia), for which astronomers have gathered convincing evidence for considering its intrinsic luminosity well established. By observing the apparent size for an extended object of which we know independently its intrinsic size, we measure its distance d (the scaling is with d in the nearby universe, from Euclidean geometry). By observing the apparent luminosity for an object of which we know independently its intrinsic luminosity, we also measure its distance (for isotropic sources, the scaling is with d^2 in the nearby universe, from the fact that the emitted power spreads out to the sphere of surface $4\pi d^2$). Of course, a number of effects may require the adoption of suitable corrections in order to make the final measurements; for example, the light from the source may suffer from absorption due to intervening material or magnification due to an intervening gravitational lens.

In practice, astronomers often make use of certain empirical scaling laws as an equivalent of standard rods or standard candles. An example of a scaling law equivalent to a standard candle is that of the empirical period–luminosity relation for the variable stars Cepheids, which allowed Hubble to measure the distance to nearby galaxies, such as M31. From a relatively easy measurement of the pulsation period (of the order of days) of a given Cepheid, by application of the period–luminosity relation we obtain a measurement of its intrinsic luminosity, which, compared to the observed apparent luminosity, yields the desired distance determination. Naturally, if we can identify in a nearby galaxy many Cepheids and carry out the measurement for each of them, we make an improved distance measurement, because, to some extent, we can limit the effects related to the fact that the period–luminosity relation is not a mathematical equation but, rather, a correlation between physical quantities with finite dispersion. A more modern example of a scaling law equivalent to a standard candle is the so-called Tully–Fisher relation for spiral galaxies, which states that

$$L = aV^4, \tag{1.1}$$

where L is the intrinsic luminosity produced by the stars of the galaxy and V its characteristic rotation speed. The quantity V can be measured by spectroscopic methods in a distance-independent way. The Tully–Fisher relation allows us to determine the intrinsic luminosity L of the galaxy and then, from the observed apparent luminosity, its distance.

As an example of standard rod we can refer to the fundamental plane relation, which applies to elliptical galaxies. Empirically it has been established that

$$\log R_e = \alpha \log \sigma_0 + \beta SB_e + \gamma, \tag{1.2}$$

where R_e is the effective radius, that is, the radius of the disk from which we receive half of the total luminosity of the elliptical galaxy (if the galaxy does not appear to us as circular but is truly elliptical, the quantity R_e can be suitably defined), σ_0 is the central velocity dispersion (a measure of the random motions of the stars in a small region close to the galaxy center), and SB_e is the mean surface brightness (in magnitudes per square arcsecond) of the galaxy inside the disk of radius R_e. Note that in the nearby universe, because for given apparent properties the intrinsic luminosity scales as d^2 and the intrinsic size scales as d, the surface brightness is distance independent. The parameters α, β, and γ are numerical coefficients that have been determined empirically. Therefore, for a given elliptical galaxy from a spectroscopic measurement of σ_0 and a photometric measurement of SB_e, the fundamental plane relation allows us to measure the intrinsic size R_e, which, compared to the apparent angular size, allows us to measure the distance to the galaxy.

Extending the concepts developed in this subsection to the study of sources at cosmological distances, that is, at finite redshifts, requires some nontrivial discussions. Quantitatively, the relevant results generally depend on the specific cosmological model that we are considering. Probably the most striking point is that the distance based on the comparison between intrinsic and apparent size is different from the distance based on the comparison between intrinsic and apparent luminosity (then we distinguish between angular-diameter distances and luminosity distances).[1] Another important effect to keep in mind is that in this context surface brightness is not distance independent, because it suffers from the so-called cosmological dimming (which depends on redshift). Another quantity related to distance that we may consider is the time passed from the emission of the photons coming from a given source to the collection of the photons by the observer. This leads to the concept of look-back time. In a monotonic way, sources at larger and larger redshifts sample conditions of the universe at larger and larger look-back times, that is, closer and closer to when the Big Bang occurred.

In this respect, a curious effect should be noted. Galaxies come in different sizes, so they should not be confused with standard rods. Yet, we may say that 10 kpc is a linear size that can be typically associated with a galaxy (e.g., the distance from the Sun to the center of our Galaxy is ≈ 8 kpc). If we plot the apparent angular size corresponding to 10 kpc as a function of redshift z (to draw such a curve we have to set the values of the main parameters that characterize the adopted cosmological model; we will briefly address the main properties of cosmological models in Section 2.2 and in Chapter 10), we find the surprising result that after a rapid Euclidean decline for $z \ll 1$ the curve reaches a minimum of ≈ 1 arcsec at $z \approx 1$, and then it is characterized by positive derivative. There are then two important surprises. The first is that beyond a certain look-back time (already reached by modern telescopes, because we observe galaxies with $z > 1$), galaxies should appear to become larger and larger. The second is that galaxies are always observed as extended sources, no matter how far they are from us.

1.3 Emission from Complex Sources and Mass-to-Light Ratios

When we observe extended sources, especially when they are far away, so that we cannot resolve their individual components, the photons that we collect are a blend of radiation emitted from a variety of sources by means of many different mechanisms. For example, the optical light from galaxies is dominated by the light emitted by stars of many different types, but also comprises light from the interstellar medium (especially in spiral galaxies, which generally host HII regions, i.e., regions of ionized hydrogen). In turn, the 21-cm observations of spiral galaxies collect the emission from a turbulent medium, made of clouds characterized by different sizes and different thermal properties, which are generally not resolved by the radiotelescope.

1.3.1 Stellar Populations

If we then refer to the stellar component of a galaxy, a small region of an image is generally associated with the emission from many many stars, including main-sequence stars, white dwarfs, neutron stars, and so on. In the visible, stars such as white dwarfs and neutron stars would contribute very little to the observed image. Much of the light may be contributed by relatively few young stars, such as O and B stars, if present, but other individually fainter stars may contribute significantly if they happen to be much more numerous. Astronomers refer to stellar populations as a term to indicate a stellar mixture with well-defined age and chemical characteristics. For spiral galaxies, Population I objects refer to the younger component, which is usually associated with a relatively thin disk. Population II objects are typically older and related to a thicker disk density distribution. In broad terms, we may think of a stellar population as a gas of stars made of stars of different types in given proportions, much like ordinary air is largely made of nitrogen, but contains other elements, such as oxygen, in given amounts.

1.3.2 Mass-to-Light Ratios

The brief digression of the previous subsection is meant to be the basis for a natural question that arises when we analyze astronomical images. At the source, how much mass corresponds to the light collected by our telescopes? In particular, for optical images of galaxies that we know are dominated by the light emitted by stars, how can we convert an observed surface brightness into a surface (projected along the line of sight) density distribution of the stars present? The key factor of conversion is called the relevant mass-to-light ratio, often denoted by M/L, for which a specification should be added to indicate for which waveband the factor is meant to be applied. Obviously, for wavebands in the visible part of the spectrum, in solar units ($M_\odot/L_\odot$) the mass-to-light ratio is expected to be of order unity, just because the Sun is not a special star. We recall that the solar mass is $1M_\odot = 1.989 \times 10^{33}$ g.

In principle, based on a number of assumptions related to the star formation rates expected in a given system, the major formation events occurred in the past (i.e., the star formation history), combined with a detailed knowledge of the results of stellar evolution, astronomers can devise population synthesis evolutionary models that may be able to predict how much mass corresponds to a given amount of emitted light by a stellar population observed at a given epoch. Color and spectral analysis of the observed light can then lead to justified estimates of the relevant mass-to-light ratio. In practice, mass-to-light ratios are best measured by means of dynamical studies (see Part II of this book).

In its simplest form, the mass-to-light ratio is a local concept. Different parts of a given galaxy may be associated with different mass-to-light ratios, because the properties of the underlying stellar populations may change from place to place. In reality, for a given galaxy or for other stellar systems, such as globular clusters, an observed homogeneity of colors and spectral features may justify the use of a constant mass-to-light ratio within the system, as a reasonable first approximation.

As we will discuss in detail in Part II, the issue of the measurement of mass-to-light ratios is strictly connected to the problem of the amount and distribution of dark matter. Especially for the cases in which we have evidence for the presence of dark matter, the concept can be defined in the sense of a cumulative quantity (by considering the ratio of the mass to the emitted light in relation to larger and larger volumes), and eventually it can be used as a global quantity; however, as we will discuss in Part II, the spatial extent of dark halos is not easy to determine empirically.

For nearby globular clusters, we can easily resolve most of their brightest stars. In these cases, we have a rather direct determination of the properties of the stellar populations that are involved. But even here the task of determining the mass-to-light ratio in a cluster and how it may change in different parts of the cluster (in particular, from the central regions to the periphery) is made difficult by the fact that white dwarfs, neutron stars, and black holes (the so-called dark remnants) are difficult to detect.

1.3.3 Doppler Shift and Doppler Broadening

If we now consider spectroscopic observations of emission from complex sources, we should be aware that the information contained in the data may be not as easy to extract as we may naively imagine. In particular, if we take spectroscopic data for a part of a galaxy (as an example, refer to the case of an elliptical galaxy), with light coming primarily from a large number of unresolved stars, we should consider the following effects. The light that we are receiving from a given piece of the galaxy comes from stars located at very different distances from the galaxy center because of projection along the line of sight (note that the star number densities and the star sizes are so small that, unless there is significant absorption from interstellar medium, the light from the most distant stars along a given line of sight proceeds directly to our telescopes, because it is not intercepted by the stars that are closer to us). Therefore, not only is the set of stars that we sample characterized by a blend of stars and star motions because we are collectively considering entire stellar populations, but the populations that contribute to the signal occupy very different regions in the galaxy that we are studying.

From the point of view of diagnosing the kinematics of the observed system, from a given small region of the galaxy we will observe an overall Doppler shift and an overall Doppler broadening of the relevant spectral lines. In general, we may argue that Doppler shifts are to be associated with mean motions present in the stellar system, whereas Doppler broadenings are to be associated largely with random motions (velocity dispersions) and gradients of mean motions present in the small region of the galaxy that we are studying. In practice, the interpretation of these data requires the specification of a suitable model. In the simplest case, by suitable model we mean a specification of the six-dimensional distribution function $f(\vec{x},\vec{v})$, that is, the probability of finding a star at a given vector position $\vec{x}$, with given vector velocity $\vec{v}$. The direct problem of deriving, from an adopted model f, the expected Doppler shifts and Doppler broadenings at a given position in the sky, by projection of the Doppler effects along the line of sight, may be technically difficult but is conceptually straightforward. In turn, the inverse problem of extracting from the observed Doppler shifts and Doppler broadenings information on the underlying model f is a difficult and generally ill-posed problem. In principle, best results could be obtained if we could measure the entire line profiles, but such measurements are extremely difficult to make. Another point that should be kept in mind and that complicates the analysis further is the fact that the effects on the line profiles are weighted by the luminosity of the individual stars that contribute to the observed emission.

For atomic hydrogen (HI) 21-cm observations of spiral galaxies, we face similar problems, which may be partly eased if the gas is distributed in a regular and symmetric thin disk. In this case, the extraction of information about the overall rotation (the rotation curve) can be made with good confidence from the observed shifts, whereas the broadening of the 21-cm line reflects the turbulent motions of the gas. In practice, a number of effects, such as the presence of noncircular motions, spiral structure, warps, asymmetries, and extra-planar gas, complicate the issue in nontrivial ways.

1.4 Dynamical Measurement of the Distance to a Globular Cluster

In this short section, we wish to introduce an interesting example of astronomical measurement based on dynamics. We have no intention to develop here realistic models of globular clusters, which would require more advanced methods and would bring us well outside the main scope of this book. To some extent, we may take the example described here as a kind of thought experiment. In reality, for some clusters for which good models are available measurements substantially similar to the one described here have already been performed.

1.4.1 The Maxwell–Boltzmann Distribution Function

From the kinetic theory of gases, we know that the particles of which a simple one-component system is made should be described by the Maxwell–Boltzmann distribution function in the six-dimensional phase space

$$f = Ae^{-\frac{E}{kT}}, \tag{1.3}$$

if the system is thermodynamically relaxed. Here E is the single-particle energy, k is Boltzmann's constant, and T the temperature of the gas. For particles of mass m, the single-particle kinetic energy is $mv^2/2$. Therefore, with respect to the velocity space, the distribution function is an isotropic Gaussian

$$f = \hat{A}e^{-av^2}, \tag{1.4}$$

where $a = m/(2kT)$. Note that $1/\sqrt{a}$ is a typical velocity that characterizes the random motions of the particles that make the gas. Mathematically, it gives a measure of the width of the Gaussian, so that a cold (hot) gas is characterized by a small (large) velocity dispersion. For a real gas, thermodynamical relaxation is ensured by the frequent collisions among the gas molecules.

1.4.2 Dynamical Distance to a Globular Cluster

It has long been debated, as briefly already described by Henri Poincaré in his essay *Science et Méthode* (1908), whether globular clusters can be considered as relaxed gases of stars. Progress in stellar dynamics and observations has led to the picture that many globular clusters should indeed be regarded, at least in their central regions, as relaxed stellar systems and thus should be characterized by a distribution function similar to that of Eq. (1.3). Without entering the issues related to the justification of this statement, let us follow here a simple consequence of astronomical interest.

Consider a relaxed globular cluster for which, in the central regions, we have collected the line-of-sight velocity measurements $\{v_{los}^{(n)}\}$ for a large number N of individual stars ($n = 1,2,\ldots,N$) and proper motions $\{\omega^{(m)}\}$ for a large number M of individual stars ($m = 1,2,\ldots,M$). The two sets of stars for which the data have been collected need not coincide.

The average values of these quantities will give us an estimate of the velocity vector that describes the three-dimensional motion of the cluster as a whole.

The dispersions around the average, σ_{los} and σ_ω, can then be measured. We recall that the intrinsic velocity dispersion $\sigma_\perp$ is related to the apparent proper motion dispersion σ_ω by the relation $\sigma_\perp = d\sigma_\omega$, where d is the distance to the cluster. From the model assumption that the velocity dispersion is isotropic, if we identify the two velocity dispersions we then obtain a measurement of the distance d. This is a dynamical measurement, because it is based on a dynamical argument. For more complex dynamical models, a dynamical measurement of the distance d can be made on the basis of the data mentioned in this simplified example.

Note

1 Distinguishing the concepts associated with the different distance definitions is quite subtle. In particular, for a given cosmological model and a given redshift, the various distances can be very different from one another. The concepts are well described in cosmology books; see also Hogg, D. W. 2000. https://arxiv.org/pdf/astro-ph/9905116.pdf. For quantitative purposes, one useful "distance calculator" is available, for example, at www.astro.ucla.edu/wright/CosmoCalc.html.

2
Optical Astronomy

Progress in astronomy is associated with the construction of new telescopes and new instruments. A systematic survey of the major optical facilities that are currently available or being planned would be soon out of date. Therefore, in opening this chapter, we will only mention a few selected initiatives of interest to give a flavor of the tools that astronomers are considering.

Similarly, on the side of science, important discoveries made in optical astronomy are so many that it would be impossible to make even a short list of key cases and to describe them even briefly. Therefore, in this chapter, we will examine only one major set of observations from space, the so-called Hubble Deep Fields, and then proceed to outline a landmark discovery made at the turn of the century, that is, the observations of distant supernovae that have led to convincing evidence that the universe is not only expanding, but, at the present epoch, is actually accelerating. Very briefly, we will also comment (Section 2.3) on one aspect of the technological progress in astrometry, because this has had a major impact on scientific issues that will be addressed in Part II, in particular on the amount and distribution of dark matter in our Galaxy.

A large investment, not only in the field of optical astronomy, is being made in placing telescopes at special locations very far from Earth. These special sites correspond to Lagrangian points, that is, equilibrium points of the restricted three-body problem for the Sun–Earth system. At the end of this chapter we will make a digression on these concepts, which will also allow us to introduce the tidal radius, a concept frequently used in astrophysical dynamics.

VLT (*Very Large Telescope*) is a set of four 8.2-m Unit Telescopes (operated by the European Southern Observatory) located at Cerro Paranal at an altitude of about 2600 m, in the Chilean Andes. First light for one of the Unit Telescopes was obtained in 1998. Another major astronomical site is the top of the volcano Mauna Kea, in one of the Hawaiian islands, where many telescopes are hosted at an altitude higher than 4000 m. A special role is played by the *W. M. Keck* Observatory, which is based on two 10-m telescopes, each of which is made of 36 hexagonal segments. The observatory has made significant advances in the use of adaptive optics and is now equipped with laser guide stars. *Keck* I started to operate in 1993; in 1996 it was joined by *Keck* II. Relevant scientific and technical information can be found at www.keckobservatory.org/ and www.keckobservatory.org/about/instrumentation.

Among the major initiatives in optical astronomy from the ground, we should mention E-ELT (*European Extremely Large Telescope*), planned to start operations in 2025; it will be placed on Cerro Amazones, at an altitude slightly above 3000 m, about 20 km from Cerro Paranal. This is a 40-m class telescope, the size of which was determined in order to fully exploit its potential for adaptive optics observations so as to reach the relevant diffraction limit. More information can be found at www.eso.org/sci/facilities/eelt/.

HST (*Hubble Space Telescope*) is a 2.4-m telescope launched in 1990 and located in a low-altitude (about 540 km) quasi-circular orbit. It has dominated the scene of optical astronomy from space. It survived an initial crisis, when just after launch it was found that the quality of the optical system was poor (one lens turned out to be out of position by more than 1 mm). Five servicing missions helped to repair and maintain the telescope and to equip it with new instrumentation. The first servicing mission in 1993 fully resolved the initial problem. The last servicing mission, originally planned for 2005, was postponed, as a result of the *Columbia* disaster in 2003, to the year 2009; it managed to replace the six gyroscopes (a key mechanical component for a proper pointing of the telescope), to repair the key instruments ACS and STIS, to install a new camera (WFC3), and a new spectrograph (COS), basically ensuring full functionality to the telescope, which might even continue operations until the year 2030 and beyond.

As a natural successor of HST, JWST (*James Webb Space Telescope*) is a 6.5-m infrared (from 0.6 to 28.5 μm) telescope that was launched in 2021. The primary mirror is made of 18 segments. It operates in the vicinity of the Lagrangian point L2 of the Sun–Earth system, with a huge sunshield protecting it from the solar heat. It carries four main instruments, including a near-infrared camera, a near-infrared spectrograph, and a mid-infrared instrument. The science mission should last 5 years, with a possible extension to a total of 10 years of operation. The main scientific goals are the study of cosmology and galaxy evolution and the formation of stars and planets. Information and updates on this major initiative can be found at https://jwst.nasa.gov/.

For an up-to-date picture of the current planning of scientific objectives in astronomy (not only optical astronomy), the reader might consult the recent document www.nationalacademies.org/our-work/decadal-survey-on-astronomy-and-astrophysics-2020-astro2020.

2.1 The Hubble Deep Fields

The Hubble Deep Fields (see www.spacetelescope.org/science/) are a set of very deep images of selected dark regions of the sky, away from the plane of the Milky Way, so as to avoid dust extinction and bright sources (also in other parts of the electromagnetic spectrum, in view of follow-up observations that might be planned in the infrared, in the radio, or in X-rays), that have allowed us to observe the faintest, most distant, and youngest galaxies ever detected. They have been used as a goldmine for a number of projects and follow-up observations that confirmed that, in these fields, we are observing how galaxies and structures formed at a time of about 1 billion years after the Big Bang, and immediately thereafter. Very recently (see http://hubblesite.org/image/4493) the Space Telescope Science Institute has released the Hubble Legacy Field, which includes observations taken by several Hubble deep-field surveys, in the wavelength range from ultraviolet to near-infrared light.

When we look at the beautiful images of the Hubble Deep Fields (e.g., see Fig. 2.1), which provide direct evidence of the richness and variety of structures formed in the visible matter in the early universe, it may be disconcerting to recall that in only relatively recent times have we come to realize that what we observe with our telescopes is embedded in very large amounts of dark matter. We actually know very little of the structure and dynamics of such dark matter, although in this respect many important issues are being clarified by modern astrophysics, as we will describe in Part II.

The first Hubble Deep Field,[1] HDF-N, was obtained in 1995 soon after the first servicing mission and addressed a field of $2.6' \times 2.6'$ in the constellation Ursa Major by means of 342 exposures (for a total exposure of more than 100 hours) taken during a period of $\approx$ 10 days with the instrument WFPC2 (the image is characterized by an incomplete square shape). The success of this observation in the northern hemisphere prompted the realization in 1998 of a corresponding deep field in the south, HDF-S, taken with a similar strategy (but it included the presence of one quasar), with the addition of observations with the instruments STIS and NICMOS.[2]

The image of HDF-N contains a few foreground stars belonging to our Galaxy and about 3000 external galaxies, mainly very young and distant objects presenting a view of the early universe. The basically similar view obtained by HDF-S confirms the overall isotropy of the universe; in addition, it offers a field

Figure 2.1 The Hubble Ultra Deep Field [credit: NASA, ESA, S. Beckwith (STScI), and the HUDF Team; see also Beckwith, S. V. W., Stiavelli, M., et al., "The Hubble ultra deep field," 2006. *Astron. J.*, **132**, 1729]. A large fraction of these galaxies are at a redshift larger than unity, some at $z > 7$, as confirmed by spectroscopic observations with large telescopes from the ground; thus some galaxies existed earlier than 1 billion years after the Big Bang. A color version of this figure is available at www.cambridge.org/bertin

for follow-up projects from the ground, for those telescopes that are located in the southern hemisphere and have no access to the region studied by HDF-N.

Thanks to the newly installed instrument ACS, with observations taken from September 2003 to January 2004, the Hubble Space Telescope obtained the Hubble Ultra Deep Field,[3] basically reaching the limit of its capabilities. HUDF consists of a total exposure time of about 11 days ($\approx$ 1 million seconds), by pointing on a small region ($200'' \times 200''$, i.e., 11 arcmin^2) of the constellation Fornax. The image captures the light of about 10,000 galaxies down to the 30th magnitude,[4] with a look-back time of 13 billion years, a few hundred million years after the Big Bang occurred (see Fig. 2.1). The HUDF was also observed with WFC3 in the ultraviolet (data released in 2014), to collect information on the processes of star formation in the very early universe.

The entire project of the Hubble Deep Fields was then completed in 2012 with the publication of the Hubble eXtreme Deep Field,[5] a combination of images taken with the instruments ACS and WFC3/IR in a period of 10 years, centered on a part of HUDF, corresponding to a total of 22 days of exposure time.

2.2 Distant Supernovae and the Acceleration of the Universe

One of the most remarkable and surprising discoveries made by optical astronomy is the measurement of the current acceleration of the universe, obtained at the end of last century.[6] Two lines of work, *The Supernova Cosmology Project* and *The High-z Supernova Search Team*, led to this extraordinary finding. In both cases, distant supernovae of type *Ia* were used as standard candles to measure the acceleration.[7]

2.2.1 Cosmological Parameters

In the standard Friedmann–Lemaître cosmological models, the overall evolution of the universe is basically determined by three quantities referred to the current epoch. These are the Hubble constant H_0, the density parameter Ω_m, and the cosmological constant Λ.

As noted earlier in Subsection 1.2.2, the Hubble constant has the dimension of a frequency (inverse time) and measures the current rate of expansion of the universe. It is often written as $H_0 = 100\ h$ km s^{-1} Mpc^{-1}; in the last few decades, many independent astronomical studies have led to the conclusion that h is in the range of 0.5 and 0.9, and in recent years the value has been converging to $h \approx 0.7$ although a more precise determination is still under debate. Note that a low value of the Hubble constant corresponds to a long distance scale and to a long time scale for the age of the universe.

The density parameter is dimensionless and proportional to the current mean matter-density of the universe ρ_0 (including dark matter), being defined as $\Omega_m = \rho_0/\rho_{crit} = 8\pi G\rho_0/(3H_0^2)$. Of course, the mean density ρ_0 refers to an average over a scale of cosmological significance, of the order of or larger than 100 Mpc. Note that the quantity $G\rho_0$ has the dimension of the square of a frequency[8] and so the adopted definition of the dimensionless density parameter Ω_m is quite natural.

Until the end of last century, the most interesting astrophysical constraints were on H_0 and Ω_m, and it was generally tacitly assumed that the cosmological constant Λ should be set to zero. In dimensionless form, the effect of the cosmological constant is described by the parameter $\Omega_\Lambda \equiv \Lambda/(3H_0^2)$.

The Einstein–de Sitter model sets $\Lambda = 0$ and $\Omega_m = 1$, with a specific relation between Hubble time ($1/H_0 = 0.98 \times 10^{10}\ h^{-1}$ yr) and age of the universe t_0, given by $H_0 t_0 = 2/3$. Note that the critical density that is associated with the Einstein–de Sitter solution is $\rho_{crit} = 1.88 \times 10^{-29}\ h^2$ g cm^{-3}.

The three quantities just defined enter the cosmological evolution equations for the universal expansion factor $a(t)$ in the following way[9]:

$$H^2 \equiv \left(\frac{\dot{a}}{a}\right)^2 = H_0^2[\Omega_m(1+z)^3 + \Omega_R(1+z)^2 + \Omega_\Lambda] \qquad (2.1)$$

and

$$\frac{\ddot{a}}{a} = H_0^2\left[\Omega_\Lambda - \frac{\Omega_m(1+z)^3}{2}\right], \qquad (2.2)$$

where $\Omega_R \equiv 1 - \Omega_m - \Omega_\Lambda$ is the curvature parameter. From these relations, we see that the contributions of the various terms scale differently with cosmic time (as traced by the redshift variable z), so that the relative ordering of gravity, curvature, and cosmological constant contributions is rapidly changing if we move backward to early epochs (i.e., to large values of the cosmological redshift z), when the gravity term is bound to dominate. In turn, from the second (acceleration) equation we see that if $\Omega_\Lambda > 0$, in the course of evolution the universe may change from a state of deceleration to a state of acceleration.

2.2.2 Distant Supernovae

The discovery resulted from a carefully designed long-term project that comprised a well-thought-through strategy to search for distant supernovae, a sound assessment that certain supernovae, in particular those of type *Ia*, can be used as standard candles, and an accurate measurement of the magnitude–redshift relation for a statistically significant sample of objects. In addition, reliable conclusions on the cosmological parameters, resulting from a discussion of the luminosity–distance relation, would require a proper estimate of undesired effects, such as those related to extinction by intervening dust.

Supernovae of type *Ia* are those that are thought to originate from the explosion of a white dwarf (a dense carbon object in the late stages of stellar evolution) in a binary system, which occurs when the white dwarf has accreted mass from its companion star to the extent of exceeding the so-called Chandrasekhar limit (of $\approx 1.4\ M_\odot$). A detailed analysis of the properties of nearby supernovae and the well-defined characteristics of the physical mechanisms that are involved provide convincing evidence that indeed these objects can be used as standard candles.

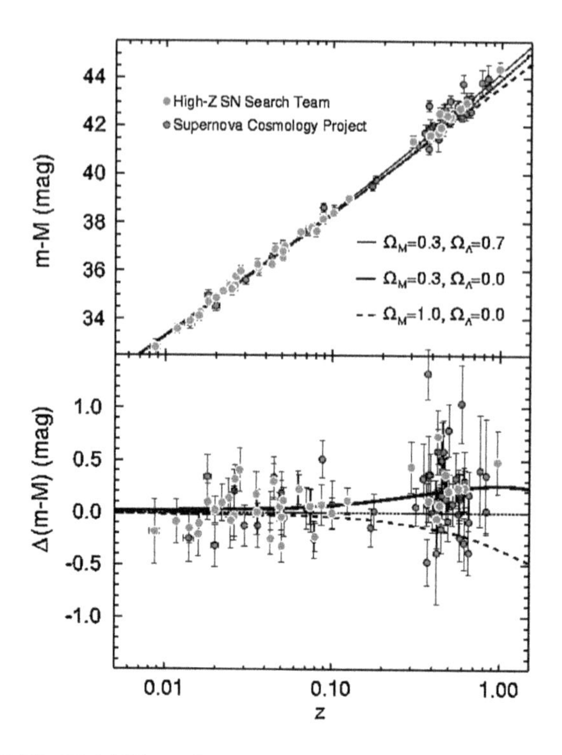

Figure 2.2 The High-Z Team discovery paper and the Hubble diagrams overplotting the supernovae measured by both teams (reprinted figure with permission from Riess, A. G., "Nobel Lecture: My path to the accelerating universe," 2012. *Rev. Mod. Phys.*, **84**, 1165. © 2012 by the American Physical Society).

The initial sample of supernovae that is associated with the discovery of the accelerated expansion of the universe consisted of 42 objects with redshift $0.18 < z < 0.83$ in the study of *The Supernova Cosmology Project* and of a comparable number of objects in the investigation of *The High-z Supernova Search Team*. This is well summarized in Fig. 2.2. Of course, since then the number of observed distant supernovae of type *Ia* has increased significantly. One interesting record case is a supernova observed at redshift $z = 1.914$, based on HST data obtained with WFC3.[10]

At present, there is growing consensus on the fact that the universe is characterized by $\Omega_m \approx 0.3$ and $\Omega_\Lambda \approx 0.7$, which would be compatible with a flat geometry ($\Omega_R = 0$). The origin of the accelerated expansion is usually interpreted in terms of a more general picture that goes under the name of dark energy. However, the data accumulated so far tend to point to the same simple picture, that of an additional constant parameter of the theory, the constant Λ, as envisaged originally by Einstein when he studied the possibility of a static, nonexpanding solution to the cosmology equations.

2.3 Astrometry: From *Hipparcos* to *Gaia*

With the launch of the *Hipparcos* mission (operating from 1989 to 1993), with astrometric precision of the order of 1 milliarcsec (mas), distances to stars measured by triangulation were extended to about 1 kpc (the final catalogue contains measurements for more than 10^5 stars).

Gaia was initially conceived as a mission called *Global Astrometric Interferometer for Astrophysics*. The mission concept evolved and changed. The astrometry mission was eventually launched at the end of 2013. It operates at the Lagrangian point L2 of the Sun–Earth system. It is meant to be the successor of the *Hipparcos* astrometry mission. As stated at http://sci.esa.int/gaia/, the main mission aims are "to measure the positions of about 1 billion stars both in our Galaxy and other members of the Local Group, with an accuracy down to 24 μas, to perform spectral and photometric measurements of all objects, to derive space velocities of the Galaxy's constituent stars using the stellar distances and motions, and to create a three-dimensional structural map of the Galaxy."[11] This practically extends the capability of measuring by triangulation the distance to most of the bright stars of our Galaxy.

The importance of the scientific achievements that are associated with the enormous progress made in the precision of astrometric measurements will be best described and appreciated in Part II. In particular, the data gathered by *Hipparcos* resolved a long-term controversy about the possible existence of significant amounts of dark matter in a thin disk in the solar neighborhood (see Section 8.2) with a clear indication that dark matter is likely to be made of collisionless matter spread out in a round halo. In Section 10.4 we will describe how *Gaia* is now leading the way to a better picture of the overall distribution of dark matter and possibly to a measurement of the global, three-dimensional total gravitational field in our Galaxy.

2.4 Lagrangian Points of the Sun–Earth System as Optimal Sites

Along the line connecting the Earth to the Sun, there are two special locations, L1 (between the Earth and the Sun) and L2 (with respect to the Earth, opposite to L1, i.e., away from the Sun), both at a distance of $\approx 1.5 \times 10^6$ km from our planet. These are two Lagrangian points of the Sun–Earth system,[12] that is, equilibrium points where a satellite can be placed and, in the frame of reference corotating with the Earth around the Sun, stay there, because gravitational and centrifugal forces properly balance. These locations are ideal for astronomical observations and transfer of astronomical data to the ground, just because they are fixed locations of the Sun–Earth system and thus offer a stationary and quiet environment from the dynamical and thermal points of view. Indeed, a number of telescopes have been (or are planned to be) positioned in the vicinity of these Lagrangian points, around which, once arrived at destination, they can orbit with minor use of fuel. Curiously, the angular size (diameter) of the Earth as seen from these Lagrangian points is approximately 0.5 degrees, that is, the same as the angular size of the Sun as seen from the Earth orbit and the angular size of the Moon as seen in our sky.

One obvious limitation of using these sites is that they are beyond the reach of currently feasible manned spaceflights: if some technical failure happens to occur, no servicing mission of the kind so successfully carried out with HST will be possible.

We have already mentioned *Gaia* and JWST in connection with L2. The L2 site will also be used by the planned *Euclid* mission (mainly devoted to the study of dark energy by means of a systematic investigation of galaxies out to $z \approx 2$). In turn, SOHO (*Solar and Heliospheric Observatory*) has been operating in a suitable halo orbit around L1 since 1996. L1 is also the location where LISA Pathfinder has been making tests since 2016, in preparation for the space mission (e)LISA (the European *Laser Interferometer Space Antenna*) that is being planned for the detection of gravitational waves. Many other missions have made or will make use of the favorable conditions offered by these Lagrangian points.

2.4.1 Tidal Forces

As a simple exercise, consider a small mass m at the end of a rope of length L, with the other end placed and kept exactly at the center of mass of a satellite orbiting on a circular orbit at distance R from the center of the Earth (assumed to be spherically symmetric). Let Ω denote the angular velocity of the satellite in its circular orbit and M_E be the mass of the Earth. Imagine the rope to be set straight, along the radial direction (i.e., along the line connecting the center of mass of the satellite with the center of the Earth). If the mass is placed below or above the center of mass, the rope will experience a small tension because the balance between centrifugal and gravitational forces is slightly violated. In formulae, the sum of centrifugal and gravitational accelerations is given by

$$a_t = \Omega^2(R+x) - \frac{GM_E}{(R+x)^2} \sim \Omega^2(R+x) - \frac{GM_E}{R^2}\left(1 - 2\frac{x}{R}\right) \sim 3\Omega^2 x. \quad (2.3)$$

The acceleration is positive (pointing outwards) if $x > 0$ (i.e., if the mass is placed above the center of mass) and negative (pointing toward the center of the Earth) if $x < 0$. In both cases, the rope will experience the tension

$$T = m|a_t| \sim 3m\Omega^2 L. \quad (2.4)$$

The tension quantifies the stretching effect of the tidal forces acting on the test mass. The symbol $\sim$ reminds us that the expression that we have obtained corresponds to a first-order expansion based on the assumption that $L \ll R$. The cancellation between the zeroth-order terms occurs because of the dynamical requirement of the circular orbit (Kepler's third law).

A variation on the previous extremely simple exercise is to consider the forces acting on a thin layer of water imagined to cover a small solid planet in circular orbit around a star. In this second exercise, we add the contribution of the gravitational attraction on the water by the planet, but eventually we come out with an estimate of the stretching due to the tidal forces that will make the water rise both in the direction of the star and in the direction opposite to it. This deformation has thus a quadrupolar character, characteristic of the tidal forces. This problem is more difficult to solve in quantitative detail. Yet, the basic solution is qualitatively suggested by the solution worked out in the simpler exercise studied above. In addition, this explains the origin of the word tidal: indeed the mechanism that we have examined briefly is at the basis of tides, as experienced by fishermen and sailors.

An even more complex variation on the same theme would be to consider the entire planet as a fluid body and to determine its shape under the action of tidal forces. This opens up an interesting line of research that has attracted the attention of scientists since the time of Newton, that is, the shape of fluid self-gravitating bodies under the action of rotation or tidal forces.[13] For fluids under the action of tidal forces, scientists generally refer to the Roche problem or the Roche ellipsoids, with important applications to the study of accretion processes in binary stars (see Chapter 4).

2.4.2 Tidal Radius

Consider a binary system made of two masses m and M, with $m \ll M$, in a circular orbit of radius R, under the action of the mutual gravitational attraction. For simplicity, assume that both objects are spherically symmetric. Let $r \ll R$ be the radius of the lighter object. Qualitatively, because the stretching tidal forces that we have examined briefly in the previous subsection are proportional to the distance from the center of mass of the object under the action of tides, if the radius r exceeded a certain value, the gravity of the light object would be unable to counteract the tidal stretching, so that equilibrium conditions in the smaller object would be impossible. This threshold value is called tidal radius (r_t). The astrophysical phenomenon related to the loss of equilibrium is often called tidal disruption. The tidal radius can thus be estimated from the condition

$$|a_t(r_t)| \sim 3\Omega^2 r_t = \frac{Gm}{r_t^2}. \tag{2.5}$$

By inserting back the expression for Ω given by Kepler's third law, we obtain

$$r_t = R\left(\frac{m}{3M}\right)^{1/3}. \tag{2.6}$$

The condition for tidal disruption is sometimes expressed in terms of a critical density ρ_t for the lighter object

$$4\pi G\rho_t = 9\Omega^2. \tag{2.7}$$

2.4.3 Lagrangian Points for the Restricted Three-Body Problem

In the classical restricted three-body problem, two point-masses, m_1 and m_2, are taken to be in a circular orbit of radius R as a result of their mutual gravitational attraction. In a frame of reference rotating with angular velocity Ω, where[14]

$$\Omega^2 = \frac{G(m_1 + m_2)}{R^3}, \tag{2.8}$$

around the z-axis perpendicular to the plane of the orbit and passing through the center of mass (taken to be at the origin of a system of Cartesian coordinates), the two masses are fixed at a proper distance from the origin and on opposite sides, along a line that we may identify with the x coordinate. Let $\vec{x}_1$ and $\vec{x}_2$ be the position vectors associated with the two masses, so that $m_1\vec{x}_1 + m_2\vec{x}_2 = 0$ and $|\vec{x}_1 - \vec{x}_2| = R$.

We now ask where a third body, of negligible mass, can be placed at rest in the (x, y) orbit plane in such a way that equilibrium occurs. The equilibrium points, that is, the locations where the gravitational and centrifugal[15] forces cancel out exactly, are called Lagrangian points.

The equilibrium condition is

$$\frac{Gm_1}{|\vec{x}_1 - \vec{x}|^3}(\vec{x}_1 - \vec{x}) + \frac{Gm_2}{|\vec{x}_2 - \vec{x}|^3}(\vec{x}_2 - \vec{x}) + \Omega^2\vec{x} = 0, \tag{2.9}$$

to be solved for $\vec{x}$ on the (x, y) plane.

In general, this problem admits five solutions $\vec{x}^{(l)}$, that is, it identifies five Lagrangian points. Three of them, labeled L1, L2, and L3, are located on the line connecting the two masses m_1 and m_2, with L1 on the segment between the two masses, and L2 and L3 outside the segment, on opposite sides. The remaining two points, L4 and L5, are located on opposite sides ($y > 0$ and $y < 0$) with respect to the line passing through the two masses; each of the two points forms, together with the locations of m_1 and m_2, an exact equilateral triangle of side R. The proof that L4 and L5, as just identified, are indeed equilibrium points is trivial. In fact, from the geometrical description we have $|\vec{x}_1 - \vec{x}| = |\vec{x}_2 - \vec{x}| = R$. Then, from the center of mass condition $m_1\vec{x}_1 + m_2\vec{x}_2 = 0$ and from Eq. (2.8) it follows that Eq. (2.9) is satisfied exactly.

When one of the two masses dominates, as is the case of the Sun–Earth system or of the Sun–Jupiter system,[16] we may set $m_1 = M \gg m = m_2$. In this case, the two points L1 and L2 are close to the smaller mass m, approximately at the same distance $\approx r_t$ from it. Point L3 then lies approximately opposite to the location of the smaller mass with respect to the larger mass M, at distance $\approx R$ from the center of mass. Note that in the case in which one of the two masses dominates, the five Lagrangian points are all located close to a circle of radius R centered on the center of mass; in other contexts this circle is called the corotation circle.

2.4.4 Stability

It can be proved that Lagrangian points L1, L2, and L3 are always unstable, whereas L4 and L5 are stable for sufficiently small values of m/M. If the mass ratio exceeds a threshold value ≈ 0.04, L4 and L5 are also unstable. An empirical proof of the stability of L4 and L5 for the Sun–Jupiter system is given by the trapping of two separate families of asteroids (called Trojan camp and Greek camp), which are found to librate in the vicinity of the corresponding Lagrangian points.

The instability of L1, L2, and L3 is rather intuitive, by considering the effect of small displacements along the line connecting the two masses. The stability of L4 and L5 is less intuitive. In any case, we should be aware that the role of rotation, and, in particular, the presence of the Coriolis force in a rotating frame, may lead to surprising (counterintuitive) results. Therefore, when rotation plays a role caution is needed in trusting purely qualitative arguments and a full linear stability analysis is generally recommended.

One instructive example for which the concepts of equilibrium and stability in the presence of rotation are usefully clarified is that of the ball, treated as a point mass, constrained to move on a spherical bowl of radius L rotating at fixed angular velocity Ω about the vertical axis, under the action of the standard gravity acceleration g. The example can be used to clarify the difference between dynamical instability and secular (dissipative) instability and the process of symmetry breaking. In the context of elementary dynamics, it presents many analogies with the more complex case of the ellipsoidal figures of equilibrium that will be briefly introduced in Section 3.5.[17]

We recall that, in its simplest form, the concept of stability refers to the limit of small (often called linear) perturbations. We then study orbits with initial conditions close to the equilibrium conditions, that is, starting from $\vec{x} = \vec{x}^{(l)} + \vec{\delta}$ and $\dot{\vec{x}} = \dot{\vec{\delta}}$, where $\vec{\delta}$ and $\dot{\vec{\delta}}$ are arbitrarily small. In the linearized limit, we may look for solutions with time dependence of the form $\exp(i\omega t)$. If some solutions are characterized by a complex ω, with the imaginary part implying an exponentially growing time dependence, the equilibrium point is unstable.

In the context of the stability of the Lagrangian points of the restricted three-body problem, a particularly simple case is the study of L1 for the case of equal masses $m_1 = m_2 = m$, for which the rotation frequency is given by $\Omega^2 = 2Gm/R^3$, where R is the distance between the two masses. If we take a Cartesian reference frame rotating at angular velocity Ω around the z-axis, we can place m_1 at $\vec{x}_1 = (-R/2, 0)$ and m_2 at $\vec{x}_2 = (R/2, 0)$, on the x-axis, so that the equilibrium point L1 is located at the origin $\vec{x}^{(L1)} = (0,0)$. In this case, the small displacement $\vec{\delta}$ is simply the vector $\vec{\delta} = (x, y)$. From Eq. (2.9), we can now write the equation of the motion for a test particle in the vicinity of L1 as

$$\ddot{\vec{\delta}} - 2\dot{\vec{\delta}} \times \vec{\Omega} - \frac{Gm}{|\vec{x}_1 - \vec{\delta}|^3}(\vec{x}_1 - \vec{\delta}) - \frac{Gm}{|\vec{x}_2 - \vec{\delta}|^3}(\vec{x}_2 - \vec{\delta}) - \Omega^2\vec{\delta} = 0, \tag{2.10}$$

which includes the presence of the Coriolis acceleration. We now note that

$$\frac{1}{|\vec{x}_1 - \vec{\delta}|^3} = \frac{1}{[(R/2 + x)^2 + y^2]^{3/2}} \sim \frac{8}{R^3}\left(1 - \frac{6x}{R}\right), \tag{2.11}$$

and

$$\frac{1}{|\vec{x}_2 - \vec{\delta}|^3} = \frac{1}{[(R/2 - x)^2 + y^2]^{3/2}} \sim \frac{8}{R^3}\left(1 + \frac{6x}{R}\right). \tag{2.12}$$

Thus, in the limit of small displacements (linear perturbations), by applying the relation $2Gm/R^3 = \Omega^2$, we can write the two components of Eq. (2.10) as

$$\ddot{x} - 2\Omega\dot{y} - 17\Omega^2 x = 0 \tag{2.13}$$

and

$$\ddot{y} + 2\Omega\dot{x} + 7\Omega^2 y = 0. \tag{2.14}$$

The combined effect of centrifugal and gravitational accelerations is repulsive along the x-axis and attractive along the y-axis. By looking for solutions with time dependence of the form $\exp(i\omega t)$, we find that nontrivial solutions are possible only if the determinant of the related linear system vanishes, that is, only if

$$\omega^4 + 6\Omega^2\omega^2 - 119\Omega^4 = 0. \tag{2.15}$$

This completes the proof that L1 is dynamically unstable. In fact, one of the two solutions for ω^2 is negative, because the product of the two solutions for ω^2 is negative; $\omega_1^2 = (-3 - 8\sqrt{2})\Omega^2$. A solution with $\omega^2 < 0$ makes it possible for some displacements to evolve in time with exponential growth, which is the signature of (linear) instability.

Notes

1 Williams, R. E., et al. 1996. *Astron. J.*, **112**, 1335.
2 Williams, R. E., et al. 2000. *Astron. J.*, **120**, 2735.
3 Beckwith, S. V. W., et al. 2006. *Astron. J.*, **132**, 1729; Stiavelli, M., Fall, S. M., Panagia, N. 2004. *Astrophys. J.*, **610**, L1.
4 As Massimo Stiavelli put it, when interviewed by CNN, this sensitivity is approximately that required to detect a firefly at the distance of the Moon.
5 Illingworth, G. D., et al. 2013. *Astrophys. J. Suppl.*, **209**, id.6.
6 Perlmutter, S., et al. 1997. *Astrophys. J.*, **483**, 565; Perlmutter, S., et al. 1998. *Nature (London)*, **391**, 51; Garnavich, P. M., et al. 1998. *Astrophys. J.*, **493**, L53; Riess, A., et al. 1998. *Astron. J.*, **116**, 1009; Perlmutter, S., et al. 1999. *Astrophys. J.*, **517**, 565.
7 To form an overall picture of the long and complex route that eventually led to such a revolutionary discovery, it may be instructive to read the account given by three leading scientists who provided a key contribution: Perlmutter, S. 2012. *Rev. Mod. Phys.*, **84**, 1127; Schmidt, B. P. 2012. *Rev. Mod. Phys.*, **84**, 1151; Riess, A. G. 2012. *Rev. Mod. Phys.*, **84**, 1165.
8 As a simple example, to introduce the meaning of the quantity $G\rho$ and the related concept of dynamical time, consider the following simple exercise. Imagine carving a hollow cylinder from north to south pole in our planet, assumed to be characterized by constant density ρ. Prove that the oscillations performed by a pebble dropped from one pole toward the center of the Earth are harmonic and estimate the relevant frequency. Calculate the time required for the pebble to go from one pole to the other and compare it with the time taken by a ballistic missile connecting the two poles by a circular orbit. For a different context, see also the definition of the critical density ρ_t for tidal disruption, given in Subsection 2.4.2.
9 For example, see Peebles, P. J. E. 1993. *Principles of Physical Cosmology.* Princeton University Press, Princeton, NJ.
10 Jones, D. O., et al. 2013. *Astrophys. J.*, **768**, id.166.
11 As an order of magnitude, 1 μas is the angle subtended by a small insect at the distance of the Moon.
12 In a simple model that applies to the vicinity of the Earth, studied in the context of the so-called restricted three-body problem, in which the Sun and the Earth are imagined to be two isolated gravitating bodies, treated as point-masses.
13 A thorough introduction to this general topic is provided by Chandrasekhar, S. 1969. *Ellipsoidal Figures of Equilibrium.* Yale University Press, New Haven, CT.
14 In the general case, the fact that the term $G(m_1 + m_2)$ is the correct constant in Kepler's third law follows from the equations describing the motion of an

equivalent particle, associated with the relative position vector, with inertial mass equal to the reduced mass $\mu = m_1 m_2/(m_1 + m_2)$, subject to a gravitational force with strength determined by the product $G m_1 m_2$.

15 Obviously, in order to identify the Lagrangian points, the Coriolis force is irrelevant, because in the discussion of the equilibrium condition the third (test) particle is assumed to be at rest.

16 For the Sun–Earth system we have $m/M = M_E/M_\odot \approx 3 \times 10^{-6}$; for the Sun–Jupiter system we have $m/M = M_J/M_\odot \approx 10^{-3}$. Note that, from Eq. (2.6), we find $r_t = R(M_E/3M_\odot)^{1/3} \approx 1.5 \times 10^8 (10^{-6})^{1/3}$ km $= 1.5 \times 10^6$ km, consistent with the statement made at the beginning of Section 2.4.

17 A discussion and solution of the simple problem of the rotating bowl can be found in Section 9.2 of the book by Bertin, G. 2014. *Dynamics of Galaxies*, 2nd ed. Cambridge University Press, New York; Chapter 10 of the same book provides a summary of some properties of the ellipsoidal figures of equilibrium.

3
Radio Astronomy

The first astronomical studies in the radio date back to the observations by Karl Jansky in 1932 (confirmed in 1938 by Grote Reber with a receiver at 160 MHz). Most likely the radio signals that Jansky correctly recognized to come from regions beyond the solar system, located in the direction of the constellation of Sagittarius, were dominated by the powerful radio source now known as SgrA*, located at the center of the Milky Way Galaxy (see Chapter 7).

The purpose of this chapter is to highlight some of the most important discoveries that radio astronomy has contributed. One of these discoveries suggests a brief comment on the interplay among density, shape, and rotation in self-gravitating systems.

A large number of radio telescopes have been built and upgraded in recent decades. Most of them are often used in cooperation as part of intercontinental networks to produce VLBI (*Very Long Baseline Interferometry*) observations, for either geodesic or astronomical studies. Here we mention only a few of them, which have played a major role in discoveries relevant to the topics addressed in Part II. In general, we may divide them into two classes, single-dish telescopes and arrays operating in interferometric mode.

The Dwingeloo Radio Observatory hosts a 25-m single-dish telescope that operated from 1956 to the year 2000; as noted in Section 3.1, this telescope led to the measurement of the rotation curve of the galaxy M31, by van de Hulst, Raimond, and van Woerden. Very large single-dish telescopes have been constructed, but generally they are fixed on the ground, that is, they are not steerable. The dish of the Arecibo telescope, operating since 1963, has a diameter of 305 m, with receivers active in the 0.3- to 10-GHz frequency range. This giant telescope has led to a number of important discoveries, in particular (see Section 3.3) the pulsar of the Crab nebula, by Lovelace in 1968, the discovery of the first binary pulsar, by Hulse and Taylor in 1974, the discovery of the first millisecond pulsar in 1982, and the discovery of the first extra-solar planets in 1992.[1] It has also been used for studies in atmospheric science and for radar astronomy, and in projects related to SETI (*Search for Extra-Terrestrial Intelligence*). It collapsed in December 2020.

WSRT (*Westerbork Synthesis Radio Telescope*, 1970) is a linear array of fourteen 25-m antennas; a major upgrade was performed at the end of the last century. The interferometer has been widely used to study the kinematics of nearby galaxies and has been the tool for decisive measurements in the discovery of halos of dark matter in spiral galaxies (see Part II). VLA (*Very Large Array*, 1980) is an array of twenty-seven 25-meter antennas arranged in a Y-shaped configuration, operating in the 0.074- to 50-GHz frequency range. Each arm of the Y-shaped configuration is 21-km long.

One of the most successful initiatives in radio astronomy has been the construction of ALMA (*Atacama Large Millimeter/submillimeter Array*), located in a desert at an altitude of about 5000 m in northern Chile. It has been in full operation since 2013. It comprises fifty-four 12-m and twelve 7-m antennas (working in the 0.3- to 9.6-mm wavelength range). With baselines from 150 m to 16 km, it can reach at its best an angular resolution of 10 mas (milliarcseconds). Extraordinary scientific results have already been obtained, especially in the field of star/planet formation and in the study of protostellar nebulae. Here we should remark that, especially after the discovery of an extra-solar planet around a solar-type star,[2] the study of protostellar and protoplanetary systems has become one of the most interesting topics at the front line of current astrophysical research.

SKA (*Square Kilometer Array*) is a project that aims at producing a giant network of antennas with total collecting area of one square kilometer. The antennas will be located in South Africa, Australia, and New Zealand, whereas the telescope headquarters are at the Jodrell Bank Observatory. Phase 1 of the project, reaching operations at the level of approximately 10 percent of the planned collecting area at low and mid frequencies (from 50 MHz to 14 GHz), has been scheduled to be completed by the year 2023. Phase 2 will lead to the operation of the entire planned collecting area at low and mid frequencies, hopefully by the year 2030. The main scientific objectives of this project are in the area of cosmology and general relativity. The scientific goals will be best approached by combining the power of SKA with complementary observations by ALMA and JWST.

3.1 The 21-cm Neutral-Hydrogen Line

A major breakthrough in radio astronomy occurred in 1944, when Hendrik van de Hulst discovered that the ground state of atomic hydrogen is a hyperfine doublet, associated with a state in which the spins of the electron and the proton are parallel and a lower-energy state in which the spins are antiparallel. The spin-flip transition energy is tiny, of $\approx 5.87433\ \mu$eV, and thus the corresponding emission is at ≈ 21.1061 cm, that is, at ≈ 1420.41 MHz. The emission line is very narrow, the mean lifetime of the excited state being about 10 Myr. Because of this, it is difficult to measure in the laboratory, but may be observed from hydrogen-rich astronomical sources.

Such a 21-cm neutral-hydrogen line was first detected in 1951 as emission from interstellar hydrogen and immediately used to estimate the rotation of the Galaxy.[3]

Neutral hydrogen (often denoted as HI), detected by means of the 21-cm emission, is ubiquitous in spiral galaxies and in other astronomical sources. It has turned out to be one of the most powerful diagnostics in studies of the dynamics of galaxies. As will be described in Part II, the study of the rotation of spiral galaxies eventually led to the discovery of dark matter halos, which brings us to the frontiers of modern astrophysics. In the following text, we briefly mention only two of the pioneering investigations in radio astronomy that the discovery of the 21-cm line has made possible.

3.1.1 Rotation Curves and Dynamics of Spiral Galaxies

HI observations of M31 made at Dwingeloo soon led to the measurement of how the disk of the Andromeda galaxy rotates around its center as a function of radius.[4] Such rotation curve (Fig. 3.1) indicates a trend at high distances from the center that, a posteriori, was recognized to carry the information that spiral galaxies are embedded in dark halos.[5]

About the same time, combined HI observations of the Milky Way Galaxy, made in Europe and in Australia, gave the first global map of the density distribution of neutral hydrogen in the Galaxy and gave interesting evidence for its spiral morphology; on the large scale, the map obtained shows features for which the average atomic hydrogen density exceeds that of one atom per cubic centimeter.[6]

3.2 Cosmic Microwave Background

One of the milestones in astrophysics, which can be called the foundation of modern cosmology, is the discovery of the Cosmic Microwave Background (CMB). An adequate account of this important discovery would deserve a book of its own. Here, we briefly summarize some key points and describe three space missions that brought us to the current picture of this phenomenon.

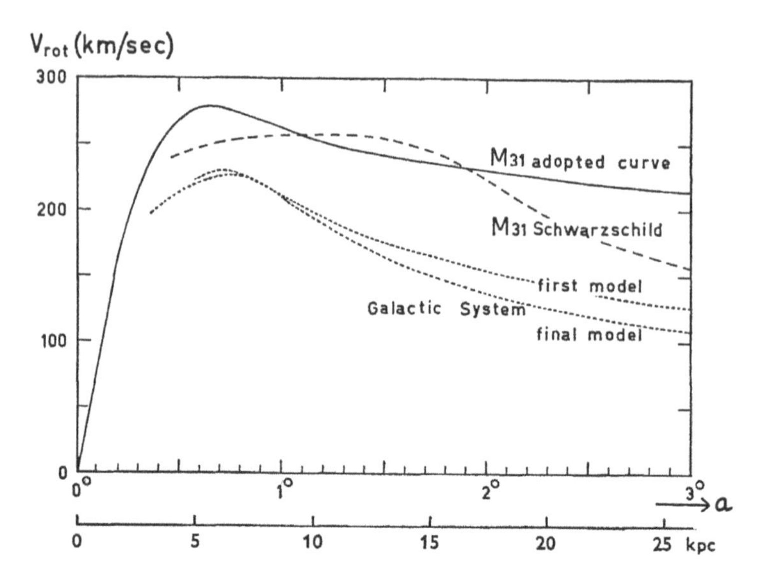

Figure 3.1 A comparison of the rotation curve for the Andromeda galaxy with that considered for the Milky Way, after the first HI observations of M31 were made at Dwingeloo (from: van de Hulst, H. C., Raimond, E., van Woerden, H., "Rotation and density distribution of the Andromeda nebula derived from observations of the 21-cm line," 1957. *Bull. Astron. Inst. Neth.*, **14**, 1). The curve marked as adopted refers to that suggested by the radio data; the excess velocity with respect to the Schwarzschild model (dashed curve) in the outer parts is related to the problem of dark matter (see Part II, in particular Chapter 9).

The study of the CMB gives us information on the state of the universe at the epoch of recombination, that is, at the epoch when the universe cooled down to produce atomic hydrogen, corresponding to a redshift $z_{rec} \approx 1100$ ($\approx 3.78 \times 10^5$ yr after the Big Bang). What we observe today is the current state of the black-body radiation that evolved basically decoupled from matter after recombination occurred (at a temperature of $\approx$ 4000 K), which we find cooled down (at $\approx$ 3 K) as a result of the cosmological expansion. Deviations from a perfectly isotropic thermal radiation give information on intervening matter and, more interestingly, on the initial stages of structure formation. In turn, from a proper modeling of the latter structures (within the concordance paradigm of the so-called ΛCDM cosmology) we gather information on the key cosmological parameters (in particular, see the parameters introduced in Subsection 2.2.1). At present, most of the attention is focused on detecting polarization signals that are predicted by inflationary models.

3.2.1 The Discovery

In 1964, Arno Penzias and Robert Wilson first detected the Cosmic Microwave Background radiation at the Bell Telephone Laboratories, as an excess signal (3.5 K) that they could not account for in terms of known radio sources.[7] Such relic of the initial Big Bang had been predicted and sought long before.[8]

3.2.2 COBE, WMAP, *Planck*

The fact that the excess radiation detected by Penzias and Wilson was to be traced back to the initial stages of the Big Bang soon became uncontroversial and recognized to be the most convincing evidence for a Big Bang origin of the universe. However, the best and most complete measurements that quantified the properties of the cosmological thermal radiation are relatively recent. A number of studies from dedicated ground, balloon, rocket, and space missions are involved. Among these, we briefly comment on three measurements from space.

COBE (*Cosmic Background Explorer*) was launched in 1989 and operated until 1993, working on a relatively low ($\approx$ 900 km) quasi-circular orbit. With data taken in 9 minutes by the FIRAS instrument, the black-body shape of the spectrum of the CMB was measured in the 0.5-mm to 1-cm wavelength range and well fitted by a Planck function with temperature $T = 2.735 \pm 0.06$ K.[9] The first year of data collected by the DMR instrument determined that the CMB is characterized by fluctuations (smoothed to a scale of 10 degrees) of 30 ± 5 μK, after removal of the dipole anisotropy; the cosmic quadrupole amplitude was measured to be 13 ± 4 μK.[10]

WMAP (*Wilkinson Microwave Anisotropy Probe*) was launched in 2001 and operated until 2010 at the Lagrangian point L2 (see Section 2.4). It produced a 13-arcmin resolution map of the CMB by means of five wavebands centered at 23, 33, 41, 61, and 94 GHz. Preliminary data and scientific results on the cosmological parameters were released in 2006, 2008, and 2010. Final maps and results, based on a 9-year data collection, were released in 2012 and published in 2013.[11] Much of the information contained in the maps is summarized in the so-called temperature–temperature angular power spectrum, which gives a quantitative description of how the deviations from isotropy are distributed as a function of the index ℓ (multipole moment) of standard spherical harmonics.[12]

The *Planck* mission basically improved, with higher resolution and sensitivity, and complemented the results obtained by WMAP. It operated from 2009 to 2013, also at the Lagrangian point L2. It carried a low frequency instrument (with three wavebands centered at 33, 44, and 70 GHz) and a high frequency instrument (with six wavebands in the range 100–857 GHz). As for WMAP, after completion of the operations, the spacecraft was placed on a heliocentric orbit, drifting away from L2. Preliminary data and scientific results were released in 2013. The final maps and results were released in 2014 and 2015. The final scientific results on the measurements of the cosmological parameters are contained in an article published in 2016.[13]

3.3 Pulsars

Neutron stars, that is, compact stars characterized by density close to that of nuclear matter, were tentatively conjectured to exist as products of supernova explosions just 2 years after the discovery of the neutron.[14] This conjecture was put forward only 3 years after Chandrasekhar discovered that nonrotating stars supported by the electron degeneracy pressure (which, especially after the work by Fowler,[15] was recognized as the proper interpretation of the stars known as white dwarfs) have a maximum mass.[16] White dwarfs are compact objects, because their mass (on the order of one solar mass) is contained in a volume similar to that of the Earth. Neutron stars are even more compact, because their mass is of the same order of magnitude, but their radius is on the order of 10 km; they are believed to be supported by the neutron degeneracy pressure. Much as white dwarfs are, neutron stars are characterized by a maximum mass.[17]

Neutron stars remained objects of theoretical speculation until in 1967 radio observations discovered a strange "pulsating" source.[18] It soon became clear that the only plausible identification for such pulsars was to be made with the neutron stars conjectured long before. It should be noted that pulsars were discovered by radio astronomers, but there are neutron stars that are not observed as pulsars. In addition, pulsars have also been observed in other wavebands at much shorter wavelengths, from the visible down to gamma rays. It has been estimated that our Galaxy hosts about 1 billion neutron stars.

The first pulsar was discovered in 1967 and is known as CP 1919, now often denoted as PSR 1919 + 21. This source was observed at 81.5 MHz and found to be characterized by a period of ≈ 1.337 s and reported to stay constant with an accuracy of 1 part in 10^7. Initially, a suggestion was made that such a regular time behavior might be associated with a pulsation of a compact star. Soon several pieces of evidence and arguments led to the picture that pulsars are to be thought of as rotating neutron stars.[19]

Some of the extended sources that are called nebulae have been recognized to be supernova remnants. In particular the Crab nebula (denoted as M1 in the Messier Catalogue) is associated with a supernova explosion that was recorded by Chinese astronomers in the year 1054. Supernova explosions may make a stellar object shine suddenly as bright as an entire galaxy. Although the explosion took place about 2 kpc away from us, the Crab supernova remained visible in daylight for about 4 weeks and at night for about 2 years; now the Crab nebula cannot be seen with the naked eye, because it is an 8.4-mag source. A pulsar associated with the Crab nebula (PSR 0531 + 21) was discovered in 1968; it is a fast pulsar, its period being of only ≈ 0.033 s.

A major discovery was made in 1982, that is, the observation of the first millisecond pulsar (PSR 1937 + 214), with a period of $\approx 1.558 \times 10^{-3}$ s.[20] As will be explained in Section 3.5, millisecond pulsars provide simple empirical evidence that pulsars are indeed stars with density similar to that of nuclear matter. The observed fast rotation has been suggested to be due to a spin-up phase during evolution associated with accretion in a binary system.

Another class of pulsars, that of binary pulsars, requires a special discussion. The term *binary pulsars* indicates pulsars in binary systems, in which the second star is a compact object, such as a white dwarf or a neutron star. Because of the fast dynamics that characterize these systems, their simplicity, and the accurate measurements that can be made by means of the clock associated with the pulsar, they can be used as fantastic laboratories for quantitative tests on the gravitational interaction, in particular on general relativity. The first and most famous case is that of the pulsar PSR 1913 + 16, discovered in 1974 at Arecibo.[21] This binary pulsar, at a distance of approximately 6.4 kpc, is made of two neutron stars of similar mass (close to 1.4 $M_\odot$) bound in a noncircular orbit (the distance between the two neutron stars is about 1.1 $R_\odot$ at periastron, about 4.8 $R_\odot$ at apoastron) of period ≈ 7.75 hours. The pulsar period is ≈ 0.059 s. As is well summarized in a relatively recent article,[22] the system has been monitored since its discovery. The data show unambiguously that the orbit is slowly decaying, with the semimajor axis shrinking at a rate of ≈ 1 cm per day, following in detail the expectations from general relativity that the phenomenon reflects the loss of energy associated with the emission of gravitational waves (Fig. 3.2); the current power associated with such emission is 0.02 $L_\odot$. A system of this kind has thus provided the first empirical evidence for the emission of gravitational waves. In some 300 million years the two neutron stars are bound to merge. This may result in a powerful event of the kind detected recently by the LIGO-*Virgo* collaboration (see a brief description in Chapter 5) and confirmed in basically all natural wavebands of astronomical observations.[23]

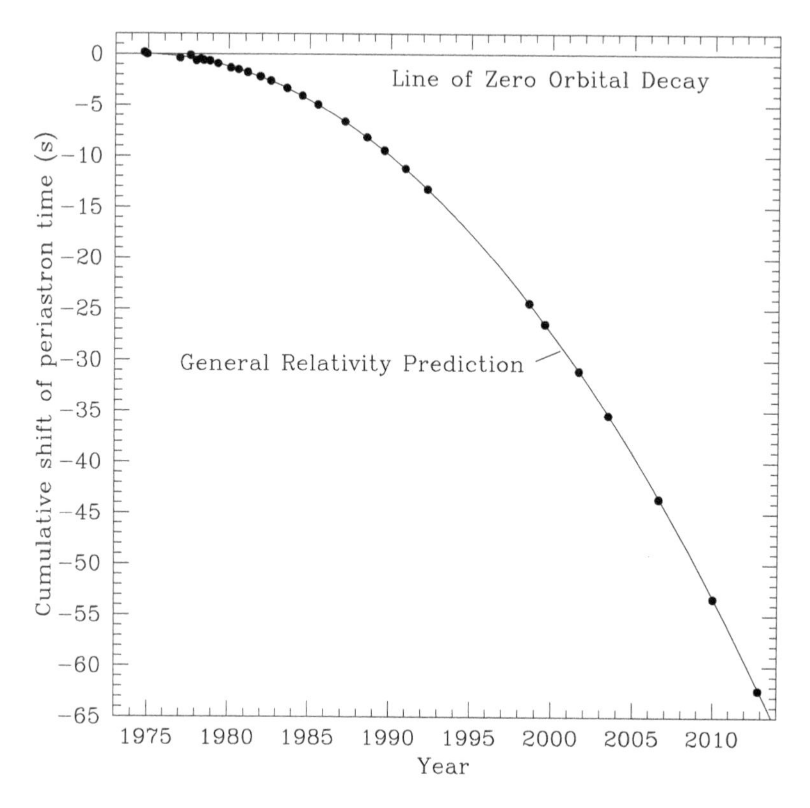

Figure 3.2 Orbital decay of PSR 1913 + 16. The data-points indicate the observed change in the epoch of periastron whereas the curve shows the expected change in epoch for a system emitting gravitational radiation, according to general relativity (from: Weisberg, J. M., Huang, Y., "Relativistic measurements from timing the binary pulsar PSR B1913+16," 2016. *Astrophys. J.*, **829**, id.55; reproduced by permission of the AAS).

3.3.1 Pulsars as Dark Remnants in Globular Clusters

In 1987, the first pulsar in a globular cluster (PSR 1821 – 24, in M28) was discovered and turned out to be a millisecond pulsar (with a period of $\approx 3 \times 10^{-3}$ s).[24] At a conference held in early 2009, it was reported that most of the known pulsars (141) in globular clusters (26 clusters) are millisecond pulsars. This fact raises interesting questions on the processes that may lead to the formation and evolution of neutron stars.

The study of the dynamics of globular clusters saw a revival of interest at the beginning of the current century. Two major hot issues for which so far we have only some important clues but no definitive answer are: (1) whether globular clusters host a central massive black hole (much like is the case for larger spheroidal stellar systems, such as bulges and elliptical galaxies; see Chapter 7), and (2) whether globular clusters contain significant amounts of dark matter.

As to the second issue, we note that, in the course of their very long evolution, the globular clusters of our Galaxy are expected to have produced large amounts of dark remnants, that is, white dwarfs, neutron stars, and stellar-mass black holes. Such remnants contribute very little to the luminosity of globular clusters. In particular, it is not easy to assess how much of the mass of the clusters, especially in the central regions, is in the form of dark remnants. The large number of pulsars detected in globular clusters confirms that dark remnants are likely to be important from the dynamical point of view. The dark matter that will be the focus of Part II is thought to be all-pervasive and is likely to be made of so far undetected exotic particles (see also Chapter 5). These considerations, combined with the fact that the amount of observed stars in globular clusters appears to be not too far from the amount required to interpret their observed kinematics, explain why it is not easy to quantify the role of dark matter in these small stellar systems.

3.4 Quasars and Gravitational Lensing

Quasars (quasi-stellar radio sources) were initially identified as strong point-like radio sources in several radio surveys of the sky. In the visible, some of these unusual radio sources were soon found to be associated with star-like objects.[25] They were recognized to be among the most distant objects in the universe; observations by the Hubble Space Telescope gave final confirmation that quasars are associated with galaxies. Quasars are now known to belong to a wide class of astrophysical sources called Active Galactic Nuclei (AGN). The term *quasar* is often replaced by the acronym QSO (Quasi-Stellar Object). In the late 1960s, it was realized that a mechanism that might explain such powerful sources over the electromagnetic spectrum is the accretion of matter on central supermassive black holes.[26] In Chapter 4, we will briefly comment on the accretion mechanism in a different context, that of X-ray binaries.

The deflection α of photons by a point mass m according to the expression $\alpha \approx 4Gm/(c^2 r)$, where G is the gravitational constant, c is the speed of light, and r is the minimum distance of a light ray passing close to the point mass, is one of the classical tests of general relativity, studied first during a solar eclipse about a century ago. It is also at the basis of the phenomenon of gravitational lensing by distributed masses. In particular, it can be shown that in the limit

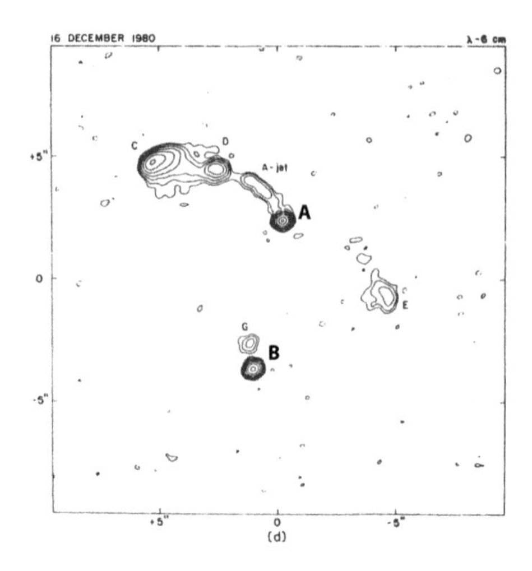

Figure 3.3 VLA radio image at 6 cm of the quasar 0957 + 561; the features denoted by A and B are the two images of a single quasar at redshift $z \approx 1.405$ produced by gravitational lensing (from: Greenfield, P. E., Roberts, D. H., Burke, B. F., "The gravitationally lensed quasar 0957+561: VLA observations and mass models," 1985. *Astrophys. J.*, **293**, 370; reproduced by permission of the AAS).

of weak gravitational fields a mass distribution associated with a mean gravitational potential Φ acts on light rays as a lens characterized by refraction index $n \sim 1 - 2\Phi/c^2$. Additional comments on this important physical phenomenon will be given in Chapter 10. Studies of many interesting phenomena associated with gravitational lensing started immediately after the formulation of general relativity.[27]

In practice, the astrophysical interest in gravitational lenses bloomed only in 1979, when combined radio and optical observations provided the first evidence for the splitting of the image of a single distant quasar as a result of gravitational lensing by an intervening galaxy.[28] The quasar, 0957 + 561, not far from NGC 3079 in the constellation Ursa Major, is at redshift $z_s \approx 1.405$ and the two images are separated by 6″ (for a VLA radio image, see Fig. 3.3). The lensing galaxy that is responsible for the double image was identified at a later stage and reported to be a giant elliptical cD galaxy at redshift $z_d \approx 0.39$.[29] The 1979 paper only argues that if the twin images are indeed produced by a lensing galaxy along the line of sight, then the galaxy should be at a redshift larger than $z_d \approx 0.1$.

Another famous example of multiple images of a single distant quasar produced by an intervening galaxy is that of the gravitational lens 2237 + 0305, known as the Einstein Cross, in which four images of a source at $z \approx 1.69$ are seen in the vicinity of the center of a lensing spiral galaxy located at $z \approx 0.039$.

Initially, much of the observational work on gravitational lenses was carried out by radio and optical observations, often in combined projects. In recent years, most studies of gravitational lenses have been performed by optical telescopes.

3.4.1 Strong and Weak Lensing

It is quite natural to expect that for a given geometrical configuration involving the observer, a gravitational lens (often called the deflector), and a distant source (all approximately aligned along a certain line of sight), the lensing effects will be stronger if the deflector is compact and massive. In addition, it is also natural to expect that if the lens is characterized by a matter density distribution ρ_d, much like in the study of thin lenses in standard optics, the decisive factor in determining the lensing effects will be the density Σ_d projected along the line of sight, associated with ρ_d. These statements can be proved and quantified, but the required calculations would go well beyond the scope of the present book.

Without derivation, it is interesting to note that, because deflections are dimensionless, to judge whether a given lens, that is, a given surface density distribution Σ_d, is expected to give rise to small or large effects, we should compare Σ_d (an intrinsic property of the lens for a specified line of sight) to a suitable reference density. It can be shown that a given geometric configuration identifies a natural reference density, which is often called the critical density,

$$\Sigma_c = \frac{c^2}{4\pi G} \frac{D_{os}}{D_{ds} D_{od}}, \tag{3.1}$$

where D_{os}, D_{ds}, and D_{od} are the distance from the observer to the source, the distance from the deflector to the source, and the distance from the observer to the deflector, respectively. The appearance in the definition of Σ_c of the speed of light c and of the gravitational constant G is quite natural in this context.

In practice, we can state that strong gravitational lensing effects, such as the generation of multiple images of a single distant source, of arcs, and (in the case of very strict alignments) of rings, are all prominent phenomena that occur when the lens is strong, that is, when Σ_d is comparable to or larger than Σ_c. In turn, when $\Sigma_d \ll \Sigma_c$ only weak distortions of the images of distant sources are produced. In the latter case, the phenomena fall within the category of weak lensing. When strong lensing is present, the modeling requires a discussion of the geometrical configuration and of the properties of the lens so as to explain the observed phenomena, which often involve only one or few distant sources. When weak lensing is under consideration, a large effort is required also in terms of an adequate statistical analysis on the very small distortions produced on a large number of sources.

As often occurs in astrophysics, a given general phenomenon, such as gravitational lensing, can be considered by astronomers in two different ways. Here, from what is known about the geometric configuration and the mass distribution associated with an intervening object (that we argue acts as a lens), we may proceed directly to prove that a given observed phenomenon is indeed the result of gravitational lensing. Conversely, we may start from the working hypothesis (possibly supported by a number of observed effects) that what we observe is due to gravitational lensing, and proceed backward from what we see, to measure the mass distribution associated with the lens; in this sense, gravitational lensing has become a powerful diagnostic tool in the general problem of determining the amount and distribution of dark matter in galaxies and in the universe (see also Part II).

Finally, we note that if we consider the typical mass properties of galaxies and clusters of galaxies that determine Σ_d, by inspection of the definition of Σ_c it is possible to realize that prominent lensing effects are more likely to occur when lenses and sources are part of the distant universe (i.e., at redshifts of cosmological interest; in this case, the proper definition of the distances mentioned in the above discussion is that of angular-diameter distances). A posteriori, this explains why the interest in gravitational lensing largely grew in only relatively recent years and is largely related to questions posed by cosmology.

3.4.2 Time Delays and Supernova Refsdal

In general, when multiple images of a single source are produced by gravitational lensing, the optical paths associated with the various images are different. Because of this, if for some reason the source is subject to a flux variation, to the observer the luminosity change will appear in the various images at different times. Indeed, in the discussion given in the 1979 discovery article for the quasar 0957 + 561 one of the items mentioned by the authors as a possible test of the origin of the two images as due to gravitational lensing is that of time delays; for that particular configuration, the authors estimate that the time delay that might be observed is "of the order of months to years." Curiously, the order of magnitude estimated here is typical of many astrophysical configurations and fits in well with realistic time scales of astronomical observations.

In this context, a particularly important application of the observation of time delays had been anticipated well before multiple images were discovered.[30] The general idea at the basis of this application is that if a time delay is measured, we obtain a measurement of the intrinsic length of the difference of light paths. This breaks the degeneracy of a geometrical configuration in which only angles and deflections are involved. As a result, in practice we have the equivalent of a "standard rod" at our disposal (see Subsection 1.2.3), and we can use this measurement to determine the distance to the deflector or, in other words, to determine the Hubble constant.

A beautiful example of this phenomenon has been studied recently and is known as supernova Refsdal. Here the distant source, with a clear flux variation, is the first supernova discovered to be multiply imaged, in 2014, by an intervening cluster of galaxies, MACS1149.5 + 2223 (at $z \approx 0.54$). The supernova exploded in one arm of a distant ($z \approx 1.49$), almost face-on spiral galaxy. Images were taken by the Hubble Space Telescope, with follow-up spectroscopic observations at VLT and *Keck*. After the supernova was discovered, in a timely article[31] it was shown that it was possible to associate the explosion with four different spots where the supernova had already appeared and also predict that the supernova should appear in a different, well-defined location in a few months.[32] After the acceptance of that article, the predicted image was discovered in HST images taken in November and December 2015.

3.5 Rotating Systems

The discovery of millisecond pulsars, described briefly in Section 3.3, is associated with a result that is very simple to derive, but serves as a direct empirical proof that pulsars, if interpreted as rotators, are indeed related to neutron stars.

In a classical, nonrelativistic model, consider a small piece of matter at the surface of a star of mass M rotating with period $T = 2\pi/\Omega$, located on the equator at distance R from the star center. In order for the piece of matter to stay in equilibrium there, the following inequality must hold:

$$g > \Omega^2 R, \tag{3.2}$$

where g represents the gravity acceleration. In words, gravity must exceed the centrifugal acceleration; the difference between gravitational and centrifugal forces is provided by the reaction of the inner part of the star, much like when we stand on the floor.

An estimate of g can be given in terms of an effective density of the star as

$$g = \frac{4\pi}{3} G \langle \rho \rangle R \approx \frac{GM}{R^2}. \tag{3.3}$$

The relation between g and effective density can be taken as the definition of $\langle \rho \rangle$. The relation between effective density (if interpreted as average density) and total mass M is only approximate, because the rotating star is likely to be flattened by rotation and, correspondingly, the relation $g = GM/R^2$ applies to only spherical objects.

By eliminating g from the two equations, we find that R cancels out and obtain a lower limit to the effective density in terms of the observed period

$$G\langle \rho \rangle > \frac{3\pi}{T^2}. \tag{3.4}$$

If we then take $T = 10^{-3}$ s, we find $\langle \rho \rangle > 1.41 \times 10^{14}$ g cm^{-3}, which is indeed close to a typical density of nuclear matter.

3.5.1 Ellipsoidal Figures of Equilibrium

The study of self-gravitating rotating models of stars has its foundation in a general problem of classical mechanics that is referred to as the problem of ellipsoidal figures of equilibrium.[33] The main goal is to determine what are the shapes that can be realized by a rotating self-gravitating body under conditions of equilibrium. Clearly, the origin of these studies goes back to the investigations by Newton on the shape of the Earth.

In the simplest case, we may consider a homogeneous (i.e., constant density ρ), incompressible, rigidly rotating (angular velocity Ω) fluid body of finite size and mass and ellipsoidal shape. If the semi-axes of the ellipsoid are called a_i, by adopting the ordering $a_1 \geq a_2 \geq a_3$, we can define two dimensionless shape parameters, that is, the polar eccentricity $e \equiv \sqrt{1 - a_3^2/a_1^2}$ and the equatorial eccentricity $\eta \equiv \sqrt{1 - a_2^2/a_1^2}$. Then the sphere is characterized by $e = \eta = 0$, an oblate ellipsoid symmetric with respect to the a_3-axis is characterized by $e \neq 0$ and $\eta = 0$, whereas the case $e = \eta \neq 0$ identifies a prolate ellipsoid symmetric with respect to the a_1-axis.

It can be shown that the gravitational potential inside a homogeneous ellipsoid is harmonic, but anisotropic, that is, in a Cartesian reference frame aligned with the principal axes of the ellipsoid, it can be written as

$$\Phi^{(int)} = -\pi G\rho \sum_{1=1}^{3} (a_i^2 - x_i^2) A_i, \tag{3.5}$$

where $A_i = A_i(e,\eta)$. The frequencies along the different axes are generally different and depend on the two eccentricities in a nontrivial manner. The proof of this statement and the calculation of the frequencies are rather difficult and will not be given here. Obviously, from Eq. (3.5) the dimensionless coefficients A_i must obey the relation $A_1 + A_2 + A_3 = 2$.

If we assume that the third axis is the rotation axis and impose, in a rotating frame of reference in which the rotating body is at rest, that the relevant potential (the sum of the gravitational and centrifugal potentials)

$$\Phi^{(rot)} = \Phi^{(int)} - \frac{1}{2}\Omega^2(x_1^2 + x_2^2) \tag{3.6}$$

be constant on the surface of the ellipsoid, we obtain two equations that involve the three dimensionless quantities $\Omega^2/(\pi G\rho)$, e, and η, which define the relevant (necessary) conditions for equilibrium. For the case of an oblate spheroid ($\eta = 0$), when the two above-mentioned equations reduce to a single equation, the final equilibrium condition has a relatively simple expression:

$$\frac{\Omega^2}{\pi G\rho} = 2\left[\frac{(3 - 2e^2)\sqrt{1 - e^2}}{e^3}\arcsin e - \frac{3(1 - e^2)}{e^2}\right]. \tag{3.7}$$

This condition relates the appropriate rotation angular velocity to a given value of the flattening of the homogeneous spheroid and defines what is commonly known as the family of Maclaurin spheroids.

From the right-hand side of Eq. (3.7), it is easily shown that the quantity $\Omega^2/(\pi G\rho)$ has a maximum value, at $e \approx 0.93$, that is, equilibrium configurations require approximately

$$\frac{\Omega^2}{\pi G\rho} < 0.449, \tag{3.8}$$

which can be written as

$$G\rho > 2.97 \times \frac{3\pi}{T^2}. \tag{3.9}$$

A comparison between Eq. (3.9) and Eq. (3.4) brings out the effects related to the flattening of the star, which were ignored in the first simple derivation.

Actually, the limit 0.449 provided by Eq. (3.8) is overestimated, because at high eccentricities the Maclaurin spheroids are unstable. At $e > e_c \approx 0.813$ there exists a family of triaxial ellipsoids (the Jacobi ellipsoids, with $\eta \neq 0$), that correspond to lower energy states and lower values of Ω. The number 0.449 should thus be replaced by the value of $\Omega^2/(\pi G\rho)$ attained at e_c, which is ≈ 0.374.

Notes

1 Wolszczan, A., Frail, D. A. 1992. *Nature (London)*, **355**, 145.
2 Mayor, M., Queloz, D. 1995. *Nature (London)*, **378**, 355.
3 Ewen, H. I., Purcell, E. M. 1951. *Nature (London)*, **168**, 356; Muller, C. A., Oort, J. H. 1951. *Nature (London)*, **168**, 357.
4 van de Hulst, H. C., Raimond, E., van Woerden, H. 1957. *Bull. Astron. Inst. Neth.*, **14**, 1.
5 See also the discussion given by Finzi, A. 1963. *Mon. Not. Roy. Astron. Soc.*, **127**, 21.
6 See Fig. 4 in the article by Oort, J. H., Kerr, F. J., Westerhout, G. 1958. *Mon. Not. Roy. Astron. Soc.*, **118**, 379.
7 See the pair of contiguous articles: Penzias, A., Wilson, R. 1965. *Astrophys. J.*, **142**, 419; Dicke, R. H., Peebles, P. J. E., Roll, P. G., Wilkinson, D. T. 1965. *Astrophys. J.*, **142**, 414.
8 In particular, see Gamow, G. 1948. *Phys. Rev.*, **74**, 505; Alpher, R. A., Herman, R. C. 1948. *Phys. Rev.*, **74**, 1737. A discussion of the historical background and developments that led to the discovery are given in many books and documents; for example, see the monograph by Peebles, P. J. E. 1993. *Principles of Physical Cosmology*. Princeton University Press, Princeton, NJ.
9 The preliminary measurement is reported by Mather, J., et al. 1990. *Astrophys. J.*, **354**, L37.
10 Smoot, G. F., et al. 1992. *Astrophys. J.*, **396**, L1.
11 Bennett, C. L., et al. 2013. *Astrophys. J. Suppl.*, **208**, id.20.
12 Figure 32 in the article just cited shows the measurements of the power spectrum in the ℓ-range 2–1100.
13 Planck Collaboration, 2016. *Astron. Astrophys.*, **594**, id.A13; the temperature–temperature power spectrum is shown in Fig. 1 of this article and extends to $\ell = 2500$.
14 Baade, W., Zwicky, F. 1934. *Phys. Rev.*, **46**, 86; Chadwick, J. 1932. *Nature (London)*, **129**, 312.
15 Fowler, R. H. 1926. *Mon. Not. Roy. Astron. Soc.*, **87**, 114.
16 Chandrasekhar, S. 1931. *Astrophys. J.*, **74**, 81.
17 The stability of these objects, in the context of general relativity, was analyzed leading to the Tolman–Oppenheimer–Volkoff mass limit (currently estimated to be $\approx 3\ M_\odot$). See Oppenheimer, J. R., Volkoff, G. M. 1939. *Phys. Rev.*, **55**, 374.
18 Hewish, A., Bell, S. J., Pilkington, J. D. H., Scott, P. F., Collins, R. A. 1968. *Nature (London)*, **217**, 709.
19 Gold, T. 1968. *Nature (London)*, **218**, 731. See also Michel, F. C. 1991. *Theory of Neutron Star Magnetospheres*. The University of Chicago Press, Chicago.
20 Backer, D. C., et al. 1982. *Nature (London)*, **300**, 615.
21 Hulse, R. A., Taylor, J. H. 1975. *Astrophys. J. Lett.* , **195**, L51. A more recent very interesting case, of two active radio pulsars in quasi-circular orbit with orbital period of $\approx$ 2.45 hours, is that of PSR J07373039A/B, monitored over a time span of more than 16 years; see Kramer, M., Stair, I. H., et al. 2021. *Phys. Rev. X*, **11**, id.041050.
22 Weisberg, J. M., Taylor, J. H. 2005. *ASP Conf. Ser.*, **328**, p. 25, eds. F. A. Rasio, I. H. Stairs. Astronomical Society of the Pacific, San Francisco, CA.
23 Abbott, B. P. et al. 2017. *Phys. Rev. Lett.*, **119**, id.161101.
24 Lyne, A. G., et al. 1987. *Nature (London)*, **328**, 399.
25 Matthews, T. A., Sandage, A. R. 1963. *Astrophys. J.*, **138**, 30.
26 Lynden-Bell, D. 1969. *Nature (London)*, **223**, 690.

27 Schneider, P., Ehlers, J., Falco, E. E. 1992. *Gravitational Lenses*. Springer-Verlag, Heidelberg.
28 Walsh, D., Carswell, R. F., Weymann, R. J. 1979. *Nature (London)*, **279**, 381.
29 Young, P. et al. 1980. *Astrophys. J.*, **241**, 507. The authors note that a proper modeling of the double image configuration by gravitational lensing requires the inclusion of the contribution by the gravitational field of the cluster of galaxies in which the cD galaxy is located.
30 Refsdal, S. 1964. *Mon. Not. Roy. Astron. Soc.*, **128**, 307.
31 In particular, see Fig. 1 in the article Treu, T. et al. 2016. *Astrophys. J.*, **817**, id.60. Many articles have been published on this subject. For a complete discussion and a comparison of different predictions and different models, the reader should consult the references cited in the article by Treu et al. and other papers that appeared shortly thereafter.
32 In the abstract of the article just cited, the authors conclude: "The future image should be approximately one-third as bright as the brightest known image (i.e., $H_{AB} \approx 25.7$ mag at peak and $H_{AB} \approx 26.7$ mag six months before peak), and thus detectable in single-orbit HST images. We will find out soon whether our predictions are correct."
33 Chandrasekhar, S. 1969. *Ellipsoidal Figures of Equilibrium*. Yale University Press, New Haven, CT.

4
X-Ray and Gamma Ray Astronomy

The study of the universe at very short wavelengths, in the domain of X-rays and gamma rays, started immediately after the beginning of the space age, with the launch in 1957 of the first artificial satellite, Sputnik 1. As noted in Section 1.1, X-rays and gamma rays are heavily absorbed by the atmosphere. With the exception of some modern tools of investigation developed in the context of the general problem of the study of cosmic rays, detectors for X-rays and gamma rays from high energy astronomical sources must operate from space. Clearly, at the beginning, scientists had no specific expectations about what kind of astronomical sources could be detected and observed. Largely under the optimistic vision of Bruno Rossi, the pioneering work carried out in the 1960s and 1970s opened up an extremely interesting, entirely new field of research, starting with the discovery of the first extra-solar source, Sco X-1, in 1962, by a rocket equipped with three Geiger counters that reached a maximum altitude of 225 km.[1]

From the 1960s to the 1970s, a transition occurred from the use of counters to the deployment of true telescopes, thanks to the construction of the so-called grazing-incidence mirrors, following a concept developed earlier by Hans Wolter. Three extraordinary discoveries mark the birth of high energy astrophysics: X-ray binaries (Section 4.1), the IntraCluster Medium (Section 4.2), and Gamma Ray Bursts (Section 4.3).

The final section of this chapter will be devoted to a brief description of the definition of fluid models. After a brief summary of the main picture, we will address the concept of hydrostatic equilibrium, which plays a key role in the study of the hot diffuse medium in clusters of galaxies and in diagnosing the amount and distribution of dark matter in those systems. We will then briefly refer to the opposite regime, of rotation-dominated inviscid disks, which is at the basis of the modeling of rotation curves in galaxies (e.g., see Section 3.1 and Part II). Finally, we will mention the general framework of the fluid description in the presence of viscosity, which is associated with the picture of accretion disks.

A few selected initiatives that have marked the progress made in X-ray astronomy should be mentioned. Some interesting studies were also performed by means of rockets and balloons; one of these led to the detection of the positron annihilation line radiation at 511 keV from the Galactic center.[2] *Uhuru* was the first satellite built with the purpose of doing X-ray astronomy. HEAO-2, or the *Einstein* Observatory, had instruments operating in the 0.2- to 3.5-keV range, with angular resolution comparable to that of some optical telescopes. Modern X-ray astronomy is associated with two initiatives. The *Chandra* X-ray Observatory (previously called AXAF) was launched in 1999 with the help of Space Shuttle *Columbia* and has been in operation since then (see http://chandra.harvard.edu). It is placed on a highly elliptic high-altitude (64-hour period) orbit. The main telescope offers a 0.5-arcsec angular resolution in the 0.1- to 10-keV energy range. The XMM-*Newton* Observatory was launched by the European Space Agency in 1999 and has also been in operation since that time. Much like *Chandra*, it is placed on a highly elliptic high-altitude (48-hour period) orbit. Similarly, the main telescope operates in the 0.1- to 12-keV energy range. Differently from *Chandra*, its best performance is not in imaging (with angular resolution of a few arcsec), but in spectroscopy.

In the domain of gamma rays, a few initiatives deserve special attention. The Vela program includes a dozen satellites launched at different times (the first in 1963, the last satellites operated until 1985) designed for military purposes, to monitor the occurrence of nuclear explosions. From 1967 on, strange Gamma Ray Bursts were detected and these were eventually shown to be of extraterrestrial origin.[3] The actual discovery that Gamma Ray Bursts are generally associated with sources at cosmological distances comes from an Italian–Dutch observatory, *Beppo*SAX (so named in honor of Giuseppe Occhialini), that operated across the turn of the century (from 1996 to 2002) in the 0.1- to 300-keV energy range. Modern gamma ray astronomy is associated with three major initiatives. The *Neil Gehrels Swift* Observatory (previously called simply *Swift*) is an observatory launched in 2004 (on a low-altitude quasi-circular orbit) designed specifically to hunt for and study Gamma Ray Bursts (in 2015 it discovered its 1000th Gamma Ray Burst). AGILE is an X-ray and gamma ray observatory of the Italian Space Agency (ASI), launched in 2007. The *Fermi* Observatory (previously called GLAST) was launched in 2008 (on a low-altitude quasi-circular orbit); it has led to a number of major surprising discoveries, among which are the first gamma ray pulsar and the so-called Fermi-bubbles of high energy radiation emitted on the large scale and associated with the Galactic center (see https://fermi.gsfc.nasa.gov).

4.1 X-Ray Binaries

The study of the first extra-solar X-ray source, Sco X-1, discovered in 1962, and of several other sources, many of them identified by *Uhuru*, soon led to the definition of a broad class of very interesting X-ray astrophysical sources, called X-ray binaries. In general, the values of the masses of the two orbiting stars are subject to significant uncertainties. A detailed classification in various subclasses is rather complex and will not be given here. To give an example of the kind of systems that are involved, we will only briefly summarize some known properties of two sources, Sco X-1 and Cyg X-1.

Sco X-1 is made of a neutron star of $\approx 1.5\ M_\odot$ and of a companion star of $\approx 0.5\ M_\odot$, with an orbital period of 18.9 hr, at a distance of ≈ 2.8 kpc. It is the (apparently) most powerful, persistent extra-solar X-ray source in the sky, with (intrinsic) X-ray power of up to $\approx 2 \times 10^{38}$ erg s^{-1} in the 2- to 10-keV energy range. It is classified as a low-mass X-ray binary.

Cyg X-1 is made of a compact object (presumably a black hole) of $\approx 15\ M_\odot$ and a companion supergiant O/B blue variable star with estimated mass of $\approx 30\ M_\odot$, with an orbital period of ≈ 5.6 days (the semimajor axis is ≈ 0.2 AU), at a distance of ≈ 1.9 kpc. The luminosity in hard X-rays, in the 40- to 200-keV energy range, is up to 10^{37} erg s^{-1}. This X-ray source was discovered in 1964 and is now considered to be the first confirmed case associated with a (stellar mass) black hole. It is classified as a high-mass X-ray binary.

The general paradigm at the basis of the interpretation of these high-energy astrophysical sources is that of accretion disks. In the accretion scenario, the emission that we observe is fueled by the material that is captured by the compact object (neutron star or black hole) from the more normal star that is orbiting in the binary system. The material is thought to accrete in the form of a relatively thin disk. In order to feed the compact object and thus release the significant amounts of gravitational energy that the inflow process makes available, the material must lose angular momentum at a rather fast rate.

This general picture was inspired by the observation of X-ray binaries and, on a much larger scale, of Active Galactic Nuclei (in particular, quasars or QSOs).[4] In practice, it revisits the framework that is thought to characterize, on a much smaller energy scale, the formation of stars and planets by means of the processes occurring in a protostellar nebula.[5]

The theory of accretion disks has thus been developed in great detail for the various contexts and has been very successful in providing a unified interpretation of a large variety of astrophysical phenomena. However, it is still based on a not fully satisfactory foundation. The observed accretion rates require an efficient angular momentum transport, higher than that expected from normal kinetic processes occurring in the disk. The accretion model thus invokes some sort of "anomalous" transport that would result from collective instabilities affecting the disk. The detailed mechanisms that lead to the anomalous transport in the various contexts still remain under debate.[6]

One interesting difference between accretion onto neutron stars and other stars and accretion disks in the presence of a central nonrotating black hole is the following. In principle, around stars the disk can reach in and be in contact with the central (compact) star, thus contributing to its rotation state and evolution. In turn, there is an inner natural boundary for a disk around a nonrotating black hole, because circular orbits are unstable at radii smaller than 3 Schwarzschild radii.[7]

4.2 Diffuse Emission from Clusters of Galaxies

The elliptical galaxy M87 in the Virgo cluster was the first extragalactic X-ray source to be detected.[8] A number of sources in the *Uhuru* Catalog turned out to be associated with clusters of galaxies. A key clue to the interpretation of the origin of the observed X-rays came from the identification of X-ray line emission from the Perseus cluster[9] and from Coma and Virgo.[10] Eventually, a detailed discussion on the observed properties of these sources, and especially of the fresh X-ray data coming from the surveys performed by the *Einstein* observatory, made it clear that, in general, clusters of galaxies are pervaded by an IntraCluster Medium (ICM), which is responsible for the diffuse X-ray emission by means of Bremsstrahlung.[11]

To give an idea of the main properties of the ICM in clusters of galaxies, we may refer to one nearby cluster (Coma; Fig. 4.1) and to a sample of ≈ 50 hot (with temperature in excess of 3.5 keV), not strongly interacting clusters that are found in the XMM-*Newton* archive in the intermediate 0.1- to 0.3-redshift range.[12]

The Coma cluster of galaxies (also known as Abell 1656) is located at ≈ 100 Mpc and contains more than 1000 galaxies. Its ICM is characterized by temperatures close to 10 keV. In the central regions of the cluster, the electron number density is estimated to be $n_0 \approx 5 \times 10^{-3}$ cm^{-3}. Inside a sphere of 0.5 Mpc, the mass associated with the hot ICM is estimated to be $\approx 0.2 \times 10^{14} M_{\odot}$, about a factor of 3–4 larger than the mass associated with the individual galaxies and about a factor of 10 smaller than the mass associated with dark matter. The issue of the amount and distribution of dark matter in clusters of galaxies will be addressed in Part II.

For the sample of ≈ 50 hot clusters at intermediate redshift mentioned above, the mean temperature T_M inside the clusters is found to cover the range 3.5–11 keV. For a given cluster, the temperature profile may exhibit the presence of

Figure 4.1 Coma cluster in X-rays (credit: X-ray: NASA/CXC/U. Chicago, I. Zhuravleva et al., Optical: SDSS). A color version of this figure is available at www.cambridge.org/bertin

a cool core (with a central temperature drop to $\approx$ 0.6–0.7 T_M) and generally declines in the outer parts (for radii larger than 300–500 kpc). The measurement of accurate and reliable temperature profiles has proven to be a very difficult task.

The ICM is often referred to as intergalactic gas. However, given its high temperature it is ionized gas, that is, plasma. It is made primarily of protons and electrons, with a significant fraction of ionized helium and small amounts of heavier nuclei. The mass density of the ICM, ρ_X, is related to the electron number density n_e and the proton mass m_p by means of the relation $\rho_X = \mu_e n_e m_p$, where a reasonable estimate for the conversion factor is $\mu_e \approx 1.17$. The equation of state for such a plasma is generally assumed to be that of a perfect gas, that is, $p_X = nkT = \rho_X kT/\mu$, with $\mu \approx 0.6$. An ideal plasma is defined as a plasma for which the number of particles in a Debye sphere is large or, alternatively, for which the interparticle Coulomb interaction is typically much smaller than the particle kinetic energy. We may define the plasma parameter[13] referred to the electron quantities as $N_D \equiv 1/g = n(4\pi\lambda_D^3/3) = 1.72 \times 10^9\ T_e^{3/2} n_e^{-1/2}$, where in the last formula T_e is meant to be expressed in eV and n_e in units of cm^{-3}. Here the Debye length is defined as $\lambda_D = \sqrt{kT_e/(4\pi n_e e^2)}$, where the subscript e refers to electrons and the electron charge is denoted by e. Then we can say that an ideal plasma is characterized by $N_D \gg 1$. If we insert for the plasma properties the numerical values quoted for the ICM in the Coma cluster of galaxies, we find $N_D \approx 2.4 \times 10^{16}$. Finally, using the same numbers, we may note that the meanfreepath for electrons in relation to Coulomb scattering is $\lambda_e = 1.44 \times 10^{12}\, T_e^2\, n_e^{-1}$ cm $\approx 3 \times 10^{22}$ cm ≈ 10 kpc. The resulting value of the electron plasma frequency $\omega_{pe} = \sqrt{4\pi n_e e^2/m_e} = 5.64 \times 10^4\, n_e^{1/2}\ \mathrm{s}^{-1} \approx 4 \times 10^3\ \mathrm{s}^{-1}$ is very low. In the formulae for λ_e and the last expression of ω_{pe}, the temperature T_e is meant to be expressed in eV and the number density n_e in units of cm^{-3}.

Under these conditions, the X-ray emission of the hot ICM plasma by Bremsstrahlung (free-free emission) is "thermal," but basically optically thin, with emissivity $\epsilon_X \sim \rho_X^2 \Lambda(T)$.[14]

The presence of the hot ICM plasma distorts the spectrum of the Cosmic Microwave Background radiation mentioned in Section 3.2 (by inverse Compton scattering; the low-energy CMB photons receive an energy boost as a result of collisions with the high-energy cluster electrons), when observed in the direction of a cluster of galaxies.[15] The phenomenon, known as the Sunyaev–Zeldovich effect, plays an important role in extragalactic astronomy.

The study of interacting clusters of galaxies (in particular, see Fig. 4.2) has led to some very important results relevant to our understanding of the nature of dark matter (see Section 10.7).

4.3 Gamma Ray Bursts

To some extent, the discovery of GRBs (Gamma Ray Bursts) presents similarities with the discovery of galaxies, which will be briefly described in Part II. In both cases, the turning point coincided with a convincing distance determination and this was immediately followed by a turning point in astrophysical research. It is also reminiscent of aspects of the search for electromagnetic counterparts for the gravitational wave events that took place in recent years and led to the beautiful discoveries associated with the gravitational wave event GW170817 (see Chapter 5).

Starting with the first clues gathered by the Vela program, it was soon realized that GRBs occur frequently (approximately on the time scale of 1 day) and with a rather isotropic distribution in the sky. The difficulty in pinpointing the direction of the sources, related to the poor angular resolution of the gamma ray observations, and thus in identifying the sources with a class of known astrophysical systems left open two completely different options for their interpretation. In one picture, the emitting sources would be very close to the Sun; in fact, the distribution of bright stars in the sky is approximately isotropic and

Figure 4.2 The Bullet cluster (1E 0657-56) in Carina, as viewed by *Chandra* and *Magellan*. The picture is ≈ 6 arcmin across. The cluster is at a distance of ≈ 3.8 billion light-years (credit: X-ray: NASA/CXC/CfA/M. Markevitch et al.; Optical: NASA/STScI; Magellan/U. Arizona/D. Clowe et al.). The relevance of this cluster to the problem of dark matter will be examined in Chapter 10. A color version of this figure is available at www.cambridge.org/bertin

does not reveal, at a first inspection, that the Sun is part of the disk of the Milky Way Galaxy. Alternatively, one may argue that the observed isotropy of the GRBs suggests that, in analogy with the distribution of galaxies on the grand scale of cosmological distances, the gamma ray explosions occur in the distant universe. Of course, in the second interpretation the observed gamma ray flux would imply that GRBs are associated with intrinsically extremely powerful explosions, most likely the most powerful events known so far.

Pros and cons were considered for the two options and a variety of models for the gamma ray emission process were developed, but the controversy remained open for about three decades. Eventually, an ingenious procedure allowed observers to resolve the puzzle. *Beppo*SAX carried instruments sensitive to a broad range of energies, from gamma rays (with poor angular resolution) to X-rays (with much better angular resolution). By switching quickly from one type of detector to another at lower energies, on the occasion of GRB 970228 it became possible to determine its position in the sky with sufficient precision[16] so as to alert observers on the ground that were finally able to identify an optical counterpart to the gamma ray explosion.[17] This led to the determination of the celestial coordinates for a GRB with unprecedented precision, and it was shown that it was associated with a faint galaxy, thus favoring the hypothesis that the explosion had taken place in the very distant universe.

After the initial breakthrough discovery, the study of GRBs has grown enormously. They appear to occur in a large variety of forms (i.e., with many different luminosity profiles and other observed properties) and they are likely to be associated with different sources. Empirically, there is convincing evidence for a bimodality in their duration time. The distribution of GRB duration presents two peaks, one below 1 second and another around 1 minute, with tails extending from 10^{-2} to 10^3–10^4 s. Furthermore, GRBs can be extremely distant[18] and extremely powerful.[19] In one case, the optical afterglow of a GRB detected by *Swift* at redshift $z = 0.937$ was visible with the naked eye for about 30 seconds (with peak visual apparent magnitude of 5.3).[20] As to the sources and the mechanisms of these powerful explosions, we would like to mention here only three points. The entire discussion of relating the observed properties to the intrinsic emission and in estimating the actual frequency of the events clearly depends crucially on whether the emission is isotropic or "directional" (jet-like). Then we note that there is, at least for a large group of GRBs, a clear association with supernova explosions. Finally, especially after the recent gravitational wave event GW170817, (some) short-duration GRBs may be related to the merging of two neutron stars.

4.4 Fluid Models

As is well known, inviscid fluids can be described by a simple set of dynamical equations. In a Eulerian description, the fluid is characterized by two scalar fields, the mass density $\rho = \rho(\vec{x},t)$ and the pressure $p = p(\vec{x},t)$, and the flow velocity vector field $\vec{u} = \vec{u}(\vec{x},t)$. For specific cases, an equation of state is generally available, which relates the pressure to the density of the fluid; if the relation is of the form $p = p(\rho)$, so that the dependence of p on the position vector is only implicit, through the dependence on ρ, then the equation of state is called barotropic. Let $\Phi(\vec{x},t)$ represent the gravitational potential responsible for the gravity forces acting on the fluid. Then, the relevant equations to be considered are the continuity equation (in the absence of sources)

$$\frac{\partial \rho}{\partial t} + \nabla \cdot (\rho \vec{u}) = 0 \tag{4.1}$$

and the Euler equation

$$\rho\left[\frac{\partial \vec{u}}{\partial t}+(\vec{u}\cdot\nabla)\vec{u}\right]=-\nabla p-\rho\nabla\Phi. \tag{4.2}$$

These equations are then supplemented by the equation of state (with the help of which we can eliminate the function p from the Euler equation) and, for gravitating fluids, by the Poisson equation

$$\nabla^2\Phi=4\pi G\rho+4\pi G\rho_{ext}, \tag{4.3}$$

where the term that includes ρ_{ext} represents the mass distribution responsible for external gravitational fields, not generated by the fluid itself (e.g., a central star or dark matter).

The continuity equation represents the conservation of mass. The Euler equation represents the momentum balance, much like $m\vec{a}=\vec{F}$ for a single particle. For viscous fluids the reference equation is the Navier–Stokes equation, which can be written as Eq. (4.2), with the addition to the right-hand side of a viscosity term (often expressed as the divergence of an appropriate viscous stress tensor). Many books introduce, justify, and illustrate the above equations, from the level of introductory physics to that of advanced fluid dynamics.[21]

4.4.1 Hydrostatic Equilibrium

We have mentioned that for the Coma cluster the visible mass is mostly contained in the ICM and that dark matter appears to be the dominant component. The statement is thought to apply to most clusters of galaxies. In this context, two major diagnostic tools are used to separate the mass contribution of the various components, as will be briefly summarized in Part II: gravitational lensing and X-ray diagnostics.

X-rays are used in a simple but rather sound and quantitative modeling process. The basic model is that of hydrostatic equilibrium. In fact, in a first approach, so far reasonably well supported by observations, clusters of galaxies appear to be generally characterized by the absence of systematic internal motions (in particular, rotation, so that, for the ICM, we may take $\vec{u}=0$). In addition, their overall morphology is frequently suggestive of quasi-stationary conditions and approximate spherical symmetry. In this framework, if we treat the ICM as an ideal, inviscid fluid characterized by density and pressure distributions ρ_X and p_X, only the radial component of Euler Eq. (4.2) would provide information and this could be written in the form

$$\frac{1}{\rho_X}\frac{dp_X}{dr}=-\frac{d\Phi}{dr}. \tag{4.4}$$

Furthermore, because of the assumed spherical symmetry, the gravitational acceleration at r can be conveniently expressed in terms of the total mass contained in the sphere of radius r, which we may call $M_{tot}(r)$, thus leading to

$$\frac{1}{\rho_X}\frac{dp_X}{dr} = -\frac{GM_{tot}(r)}{r^2}. \tag{4.5}$$

In practice, the right-hand side of the last equation is an explicit solution of Poisson Eq. (4.3). We can also consider the ideal equation of state mentioned in Section 4.2

$$p_X = nkT = \frac{1}{\mu}\rho_X kT, \tag{4.6}$$

and separate, on the right-hand side of Eq. (4.5), the contribution of the ICM from that of galaxies and dark matter, as

$$M_{tot}(r) = 4\pi \int_0^r x^2 \rho_X(x)dx + M_{gal}(r) + M_{DM}(r). \tag{4.7}$$

Here the last two terms represent the contribution to the total mass by galaxies and dark matter, respectively:

$$M_{gal}(r) + M_{DM}(r) = 4\pi \int_0^r x^2 \rho_{ext}(x)dx, \tag{4.8}$$

with $\rho_{ext} = \rho_{gal} + \rho_{DM}$. In conclusion, we obtain

$$\frac{1}{\mu\rho_X}\frac{d(\rho_X kT)}{dr} + \frac{4\pi G \int_0^r x^2 \rho_X(x)dx}{r^2} + \frac{GM_{gal}(r)}{r^2} = -\frac{GM_{DM}(r)}{r^2}. \tag{4.9}$$

X-ray photometric and spectroscopic observations (from the known properties of Bremsstrahlung optically thin radiation and the assumed spherical symmetry) provide information on the ICM properties ρ_X and T and its mass contribution $4\pi \int_0^r x^2 \rho_X(x)dx$, whereas optical observations give information on the galaxy distribution and its contribution $M_{gal}(r)$. Therefore, with the use of these data, the hydrostatic equilibrium condition Eq. (4.5) eventually determines the amount and distribution of dark matter present in the cluster.

4.4.2 Rotation-Dominated Inviscid Fluid Disks

If we now refer to the rotation curves of spiral galaxies as measured by 21-cm radio observations (see Section 3.1), we may consider a simple quasi-stationary, inviscid fluid model of a thin disk (the HI layer) dominated by rotation. Under the assumption of axisymmetry, suggested as the natural basic state for many galaxies (in which, of course, spiral arms are treated as a perturbation to be studied separately), the rotation of the disk on the equatorial plane can be described by the velocity field

$$\vec{u} = V(r)\vec{e}_\theta. \tag{4.10}$$

Here the function $V = V(r)$ is generally called the rotation curve; in addition, the rotation is generally differential, in the sense that the function $\Omega \equiv V/r$ is not constant, as it would be for the case of rigid rotation.

For such an axisymmetric fluid disk, on its equatorial plane the radial component of Euler Eq. (4.2) reduces to

$$\frac{V^2}{r} = \frac{1}{\rho_{HI}}\frac{dp_{HI}}{dr} + \frac{d\Phi}{dr}, \tag{4.11}$$

where ρ_{HI} and p_{HI} represent density and pressure for the fluid model of the atomic hydrogen layer studied by the radio observations. Two important points to be kept in mind that will be further discussed in Part II are the following. For galaxy disks the pressure gradient term is usually negligible, so that Eq. (4.11) becomes identical to the momentum balance equation for circular orbits of a single particle. In addition, differently from the case of spherical symmetry, for galaxy disks there are no simple explicit expressions that relate the distribution of visible mass in the disk to its contribution to the gravitational acceleration $d\Phi/dr$ in the plane.

4.4.3 Accretion Disks

Without pretending to formulate the key features of the fluid model in the theory of accretion disks, we wish to take the opportunity of the dynamical digression of Section 4.4 to make a simple remark that may serve to clarify some statements made in Section 4.1, when we briefly described the discovery and some properties of X-ray binaries.

For an axisymmetric, rotation-dominated, nonstationary, thin fluid disk, with surface density distribution $\sigma(r)$ and velocity field $\vec{u} = u_r\vec{e}_r + r\Omega(r)\vec{e}_\theta$, it can be shown that the angular-momentum balance equation (that is derived from the Navier–Stokes equation) can be written as

$$\frac{\partial}{\partial t}\left(\sigma r^2\Omega\right) + \frac{1}{r}\frac{\partial}{\partial r}\left(r^3\sigma\Omega u_r\right) = \frac{1}{r}\frac{\partial}{\partial r}\left[\nu\sigma r^3\left(\frac{d\Omega}{dr}\right)\right], \tag{4.12}$$

where ν represents the kinematic viscosity. The right-hand side describes the result of the viscous torque, which would vanish in the absence of shear (i.e., for rigid rotation $d\Omega/dr = 0$). In turn, the left-hand side is made of the two contributions (time-dependence and radial convection) that determine the change of angular momentum associated with such a torque. Here we wish to note how this equation summarizes the mutual interaction of three factors, that is, time dependence (the first term on the left), radial accretion (the second term on the left), and viscosity (the term on the right), which are all ignored in the simple quasi-stationary inviscid fluid model mentioned in the previous subsection.

Notes

1 Giacconi, R., Gursky, H., Paolini, F. R., Rossi, B. B. 1962. *Phys. Rev. Lett.*, **9**, 439.
2 Leventhal, M., MacCallum, C. J., Stang, P. D. 1978. *Astrophys. J. Lett.*, **225**, L11.
3 Klebesadel, R. W., Strong, I. B., Olson, R. A. 1973. *Astrophys. J. Lett.*, **182**, L85.
4 See Prendergast, K. H., Burbidge, G. R. 1968. *Astrophys. J. Lett.*, **151**, L83; Lynden-Bell, D. 1969. *Nature (London)*, **223**, 690.
5 As envisaged long ago by von Weizsäcker, C. F. 1948. *Z. Naturforsch.*, **3A**, 524; Lüst, R. 1952. *Z. Naturforsch.*, **7A**, 87.
6 A general introduction to the dynamics of accretion disks is given by Pringle, J. E. 1981. *Annu. Rev. Astron. Astrophys.*, **19**, 137. There are several excellent books on this general topic; one of these is by Frank, J., King, A., Raine, D. 2002. *Accretion Power in Astrophysics*. Cambridge University Press, Cambridge, UK.
7 The location of the innermost stable circular orbit at $r_{ISCO} = 6GM_{BH}/c^2 = 3r_S$ can be calculated in Newtonian dynamics by showing that the curvature of the effective potential per unit mass around a black hole of mass M_{BH} changes sign at $r_{circ} = 3r_S$; for $r > 3r_S$ the curvature identifies the frequency of the so-called epicyclic oscillations (see Section 7.5). The calculation can be carried out in a straightforward way by approximating in a quasi-Newtonian fashion the potential of the black hole as $\Phi_{PW} = -GM_{BH}/(r - r_S)$; see Paczyńsky, B., Wiita, P. J. 1980. *Astron. Astrophys.*, **88**, 23.
8 Byram, E. T., Chubb, T. A., Friedman, H. 1966. *Science*, **152**, 66; Bradt, H., Mayer, W., et al. 1967. *Astrophys. J. Lett.*, **161**, L1.
9 Mitchell, R. J., Culhane, J. L., et al. 1976. *Mon. Not. Roy. Astron. Soc.*, **176**, 29p.
10 Serlemitsos, P. J., Smith, B. W., et al. 1977. *Astrophys. J. Lett.*, **211**, L63.
11 Sarazin, C. L. 1988. *X-Ray Emission from Clusters of Galaxies*. Cambridge University Press, New York.
12 Leccardi, A., Molendi, S. 2008. *Astron. Astrophys.*, **486**, 359.
13 www.nrl.navy.mil/ppd/content/nrl-plasma-formulary
14 For a thorough discussion, see Sarazin, C. L. 1988. *X-Ray Emission from Clusters of Galaxies*. Cambridge University Press, New York.
15 Weymann, R. 1966. *Astrophys. J.*, **145**, 560; Zeldovich, Ya. B., Sunyaev, R. A. 1969. *Astrophys. Sp. Sci.*, **4**, 301; Sunyaev, R. A., Zeldovich, Ya. B. 1972. *Comments Astrophys. Sp. Phys.*, **4**, 173; Rephaeli, Y. 1995. *Annu. Rev. Astron. Astrophys.*, **33**, 541.
16 Costa, E., Frontera, F., et al. 1997. *Nature (London)*, **387**, 783 report the discovery of an "X-ray afterglow" associated with the GRB of February 28, 1997.
17 van Paradijs, J., Groot, P. J., et al. 1997. *Nature (London)*, **386**, 686. The fading optical counterpart was identified less than 21 hr after the Gamma Ray Burst.
18 For example, GRB 090429B may be at redshift $z = 9.4$, as shown by Cucchiara, A., Levan, A. J., et al. 2011. *Astrophys. J.*, **736**, 7.
19 GRB 080916C was observed by the *Fermi* team, with estimated redshift $z \approx 4.35$ and an apparent (under the assumption of isotropic emission) energy release of 8.8×10^{54} erg; see Abdo, A. A., Ackermann, M., et al. 2009. *Science*, **323**, 1688.

20 GRB 080319B; see Bloom, J. S., Perley, D. A., et al. 2009. *Astrophys. J.*, **691**, 723.
21 An elementary but thorough and lively introduction can be found in Feynman, R. P., Leighton, R. B., Sands, M. 1963. *The Feynman Lectures on Physics.* Addison-Wesley, Reading, MA. A more advanced reference book is Batchelor, G. K. 1967. *Fluid Dynamics.* Cambridge University Press, Cambridge, UK. A short introduction to the two alternative approaches of a Eulerian description or a Lagrangian description of small perturbations in fluid dynamics is available, for example, in Section 13 of Chandrasekhar, S. 1969. *Ellipsoidal Figures of Equilibrium.* Yale University Press, New Haven, CT, and in Section 15 of Drazin, P. G., Reid, W. H. 1981. *Hydrodynamic Stability.* Cambridge University Press, Cambridge, UK.

5

Astroparticle Physics, Gravitational Waves, and Space Physics

Astroparticle physics is a relatively modern expression that denotes research at the interface between particle physics and astrophysics.[1] In reality, this general area of research has very old origins that are rooted in the study of stellar structure and cosmic rays.

As far back as in the nineteenth century, scientists started to wonder what could be the source of the energy of the Sun and the stars, given the fact that the natural time scale associated with gravitational energy and thermal energy (what is often referred to as the Kelvin–Helmholtz time scale) is too short. Note that if we insert the relevant numerical values in the expression

$$\tau_{KH} \equiv \frac{GM_\odot^2}{R_\odot L_\odot}, \tag{5.1}$$

where we have used standard notation for the radius, mass, and luminosity of the Sun, we find $\tau_{KH} \approx 10^{15}$ s $\approx 3 \times 10^7$ yr. So defined, the time scale refers to the gravitational energy; the virial theorem (see Chapter 6) guarantees that the time scale referred to the thermal energy is of the same order of magnitude. The time scale relative to the consumption of chemical energy is even shorter.

The idea that starlight is ultimately due to nuclear reactions is generally associated with the work of Arthur Eddington in the 1920s.[2] Great strides were made thereafter. The study of nuclear astrophysics soon led to the discovery that some weakly shining stars (such as Sirius B, a companion of Sirius, the brightest star in the visible sky), called white dwarfs, are stars for which equilibrium against gravitational collapse is supported by the degeneracy pressure of the electrons.[3] It was also realized that there is an upper value of the white dwarf mass, the Chandrasekhar limit (currently estimated to be $\approx 1.4\ M_\odot$), above which equilibrium cannot be supported.[4] Neutron stars were eventually discovered to exist only about 30 years after they were conjectured, when pulsars were first detected by radioastronomers (see Chapter 3).

Under the category of nuclear astrophysics we may mention a number of currently active research lines. In addition to the study of the structure and evolution of normal stars (scientists are trying to determine empirically accurate cross sections for some low-energy nuclear reactions) and of compact stars (white dwarfs and neutron stars), we may refer to the key studies that address the modeling of supernova explosions and the nucleosynthesis processes that occur in such extraordinary events. It has long been realized that in the initial stages of the Big Bang only light elements could be synthesized, so that the heavy elements that we see must have been produced at a later stage. Supernovae have been recognized to be the natural "chemical" (nuclear) factories for the purpose.

Another research area of very long tradition is the study of cosmic rays. In 1912 balloon experiments carried out by Victor Hess provided convincing evidence for the existence of incoming radiation of extraterrestrial origin.[5] Currently, cosmic rays generally denote the component of high-energy massive particles of extraterrestrial origin, largely dominated by protons and atomic nuclei, whereas gamma rays (photons) and extraterrestrial neutrinos are considered separately. The initial studies of cosmic rays played a key role in particle physics. To mention some milestones in the field, the positron was discovered in 1932 and the muon in 1936 on the basis of studies of cosmic radiation. Cosmic rays can have energies much higher than those achieved in the most powerful accelerators currently available. Indeed an entire research line addresses the issue of the mechanisms that are able to accelerate cosmic rays to the huge energies that are observed. Many mechanisms have been proposed and others are still under investigation.[6] One theoretical prediction, the GZK limit, sets an interesting scale for the maximum energy of cosmic rays.[7] Curiously, the highest energy cosmic ray detected so far appears to be characterized by an energy of $\approx 3 \times 10^{20}$ eV, which is of the same order as the GZK limit.

Then, as is well known, a by now classical topic in astroparticle physics is the identification of the constituents of dark matter, if imagined to be made of exotic particles (in particular, WIMPs, Weakly Interacting Massive Particles). These investigations comprise "experiments" such as DAMA.[8]

As a major discovery representative of the beauty of astroparticle physics, in Section 5.1 we will briefly describe the origin and the solution of the solar neutrino problem.

Gravitational wave astronomy represents a new astronomical window, just opened. The search of gravitational waves was initially carried out by means of mechanical detectors, that is, of suspended massive bars or spheres, that were hoped to respond to signals of sufficiently high amplitudes and relatively high frequencies.[9] Eventually, great efforts went into major initiatives based on the concept of laser interferometry. In the United States, LIGO (*Laser Interferometer Gravitational-Wave Observatory*) was built and started to operate in 2002. In Europe, the *Virgo* project was approved in 1994, developed, and completed in the year 2003, especially as the result of the enthusiastic contribution of Adalberto Giazotto and Alain Brillet. Plans are under way to set an interferometric detector in space, exploiting the advantages of the enormous baselines that are available on the scale of the solar system, suited for the detection of low frequency signals (such as those expected from close binaries and from binary supermassive black holes in galaxies); the mission (e)LISA is currently scheduled to be launched in the 2030s, while the mission LISA Pathfinder has already obtained encouraging results (see also Section 2.4).

Space physics may appear to be a field of research fully separate from astroparticle physics and gravitational waves. In reality, the unifying power of astrophysics manifests even in this context. On the one hand, space missions have been considered recently in order to test alternative theories of gravity (see Part II), as a possible resolution of the problem of dark matter, for which astroparticle physics gives an interpretation in terms of so far undetected particles. On the other hand, the next-generation detectors of gravitational waves will probably work from space (in particular, the planned mission (e)LISA).

Space physics has the distinctive feature of being at the interface between physics and astrophysics. In fact, *in situ* measurements in various regions of the solar system have allowed us to collect data that are very similar to those obtained in laboratory experiments. In contrast, the sources of astronomical observations are so remote that they escape any direct assessment by the astronomer of the physical parameters that are involved.

Interplanetary probes have allowed us to make direct measurements of planetary structures and atmospheres, to study collective phenomena in planets and planetary rings, to probe the properties of planets, satellites, asteroids, and comets with imaging, gravity, and chemical analyses carried out on site. Many curious phenomena have been discovered or inspected from very short distances. We have observed hurricanes in planets and monitored from close by the fantastic evolving structures induced by shearing flows in the outer layers of Jupiter and Saturn. As a particularly beautiful example of a mysterious phenomenon that has attracted attention in recent decades, we should mention the discovery of the extremely regular and persistent hexagon at Saturn's north pole, first noted by Voyager and then revisited by the *Cassini–Huygens* mission.[10] The length of the sides of the hexagon is $\approx 15 \times 10^3$ km.

One important area of research of space physics[11] is that of space plasmas. A central focus of interest of these studies is the structure and dynamics of the solar wind. Probably the first discussions on the existence of a solar wind, that is, of streams of particles flowing out of the Sun, resulted from the first observations of solar flares in the nineteenth century and their connection to geomagnetic storms. Later, in the first part of the twentieth century, the concept of solar wind became related also to the observed properties of the solar corona during eclipses and to those of comet tails. Finally, Eugene Parker developed a dynamical picture of the heliosphere, thus providing a global framework for its interpretation.[12] This theoretical advance took place precisely at the opening of the space era when, in 1959, Soviet space missions led to a direct detection of the solar wind. Since then, many space missions have studied the properties of the solar wind by means of direct measurements in space.[13] Among the many topics of specific interest, we should mention the study of the interaction of the solar wind with planetary magnetospheres, the origin of polar aurorae (also observed in other planets), the dynamics of the Earth's magnetotail, and the properties of the equatorial current sheet and the magnetic sector structure in the heliosphere.

The Voyager interplanetary missions carry instruments for specific plasma measurements, some of which are still functioning and sending data more than 40 years after launch (1977); the two probes are currently exploring "the outermost edge of the Sun's domain," toward the boundary with the interstellar space. Among the many space missions designed to perform plasma measurements, we should mention *Ulysses*, launched in 1990, which ended its operations in 2009. *Ulysses* managed to study the properties of the solar wind at all latitudes, by performing an out-of-the-ecliptic orbit. The *Parker* Solar Probe was launched in 2018 with a number of scientific objectives, among which is the *in situ* study of the properties of the heliosphere at a distance of about 10 solar radii from the Sun. Finally, as a piece of curiosity related to concepts that are under investigation even in view of possible applications to interstellar travel, we wish to mention IKAROS (*Interplanetary Kite-craft Accelerated by Radiation Of the Sun*), launched by Japan in 2010, the first probe to use radiation pressure from the Sun (a solar sail) as its main propulsion.

As a main focus of interest in space physics, in Section 5.3 we will briefly describe one property of the solar wind.

The dynamical digression at the end of this chapter is prompted by a well-known effect in celestial mechanics that has played a key role in interplanetary navigation (in particular, for the *Ulysses* mission), that is, the so-called slingshot effect. The digression offers an excuse for mentioning a hot topic on the grand scale of galaxies, in which gravity assists can produce hypervelocity stars, an interesting phenomenon currently under investigation.

5.1 The Solar Neutrino Problem

A thorough account of the solar neutrino problem is given in the monograph by John Bahcall, which appeared just before the problem was finally resolved.[14] The problem was raised by the first measurements of the solar neutrino flux (the Homestake Solar Neutrino Detector started to operate in 1970), which exhibited a clear discrepancy with respect to the flux predicted by the solar standard model.

The solar standard model is centered on the picture that the fusion energy associated with the observed solar luminosity is basically related to a relatively simple set of reactions called the pp chain, with a small contribution from a different set of reactions that are called the CNO cycle. Eventually the reactions of the pp chain convert four protons into alpha particles, with emission of electron neutrinos of various energies. From the energy balance point of view, the basic process is associated with an energy yield of 25 MeV corresponding to the mass difference between the four protons and the alpha particle, that is, an efficiency of $\epsilon \approx 0.7\%$. In the neutrino spectrum for the standard model, the highest energy neutrinos (with energies exceeding 10 MeV) are those produced by one reaction, which involves the decay of ^{8}B, with a small contribution by the so-called hep reaction (the fusion reaction of a proton with ^{3}He); the overall flux (more than 10^{10} neutrinos per square centimeter per second at the location of the Earth) is dominated by the fusion of two protons into ^{2}H, associated with the emission of low-energy neutrinos (below 1 Mev).

Note that the time scale associated with such nuclear burning can be estimated as

$$\tau_{nuc} \approx \epsilon \times 0.1 \times \frac{M_\odot c^2}{L_\odot}, \tag{5.2}$$

where c is the speed of light and the factor 0.1 represents the fraction of the solar mass that is used during the life of the star on the Main Sequence, following the fusion reactions briefly described above. The resulting value $\tau_{nuc} \approx 10^{10}$ yr is then consistent with the age of the Sun ($\approx 4.6 \times 10^9$ yr, determined by geological studies). The photon luminosity of the Sun[15] is $L_\odot = 3.86 \times 10^{33}$ erg s^{-1}. In the standard solar model the neutrino luminosity is $0.023\ L_\odot$, the central density is 148 g cm^{-3}, and the central temperature is 15.6×10^6 K; the peak of energy production occurs at $0.09 R_\odot$, where density and temperature have declined to 95 g cm^{-3} and 14×10^6 K, respectively. We recall that the average density is 1.41 g cm^{-3} whereas the effective surface temperature is 5.78×10^3 K. Optical observations of the Sun refer to its surface and thus reflect the current state of the solar interior with a slight delay (because the typical time scale for a photon to propagate from the center to the surface, through a random walk

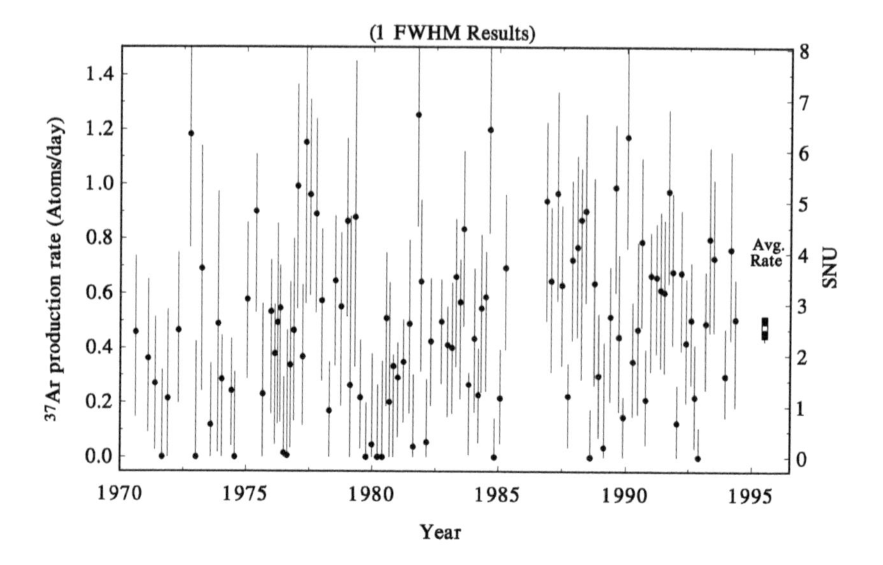

Figure 5.1 The solar neutrino discrepancy. About three decades of data collection are summarized in this plot, showing the results for 108 individual solar neutrino observations made with the Homestake chlorine detector (from: Cleveland, B. T., Daily, T., et al., "Measurement of the Solar Electron Neutrino Flux with the Homestake Chlorine Detector," 1998. *Astrophys. J.*, **496**, 505; reproduced by permission of the AAS). The prediction of the standard model would be 7.9 SNU.

process, is $\tau_{ph} \approx 10^4$ yr). In turn, measurements of the solar neutrino flux give a basically instantaneous picture of the solar interior.

The Homestake Experiment was sensitive to solar neutrinos interacting with the chlorine nuclei of a huge underground tank filled with C_2Cl_4, following the reaction $\nu_e + {}^{37}\text{Cl} \rightarrow e^- + {}^{37}\text{Ar}$, which is characterized by a minimum neutrino energy threshold of 0.814 MeV; it was mostly sensitive to the neutrinos produced by the decay of ^{8}B and ^{7}Be in the pp chain. The detector measured the solar neutrino flux by means of a careful counting of the argon atoms produced by this reaction. The measurements indicated a flux $\approx 1/3$ of that expected from the solar standard model (Fig. 5.1).

The Japanese *Kamiokande* II experiment, operating in the period 1985–1990, considered detections associated with the Čerenkov light emitted by electrons scattered by incoming neutrinos. It was best suited to measure neutrinos from the ^{8}B decay (with a threshold neutrino energy of 9 MeV) and indicated a flux slightly below 1/2 of that expected from the standard solar model, with the added decisive feature of a good correlation with the direction of the Sun.[16] A detector based on a similar strategy, but not suited to capture the low-energy neutrinos from the Sun, was IMB, the Irvine–Michigan–Brookhaven detector.

The discrepancy between observed neutrino flux and expected neutrino flux opened the way to a controversy about its origin. Broadly speaking, we may try to explain such a situation by arguing that something is wrong in the standard solar model, which thus would require substantial revision. As possible items to be considered in such revision, we may mention the role of anomalous diffusion, different types of collective instabilities, differences in chemical composition of the solar interior, exceedingly large rotation or magnetic fields, or the presence of non-Maxwellian distribution functions. Curiously, some proposed explanations were based on the existence of a small central black hole and others on the role of WIMPs, the weakly interacting massive particles that are currently thought to be the constituents of dark matter. Many of the mechanisms thus proposed were basically excluded by the study of the natural oscillations of the Sun (a topic that is often referred to as helioseismology). To complicate the matter further, at some point a correlation was claimed between the measured neutrino flux and the 11-year cycle of sunspot magnetic activity. Eventually, the standard model remained robustly untouched and the explanation turned out to be related to the basic physics of neutrinos. Initially some attention was given to the possibility of vacuum oscillations of the neutrinos, in line with earlier thoughts by Bruno Pontecorvo.[17] The key mechanism was then identified in the MSW effect,[18] a kind of neutrino oscillation that occurs in the presence of matter (especially inside the Sun). The gist of the argument is that neutrinos are massive particles and electron neutrinos can transform into muon or tau neutrinos during propagation, which then escape detection in the experiments (sensitive to electron neutrinos). In the MSW effect, the matter in the Sun enhances the probability that an electron neutrino will oscillate into a neutrino of a different type.

For completeness, we should mention that a comprehensive measurement of the solar neutrinos associated with the pp chain was finally obtained by *Borexino*, an underground experiment at Gran Sasso.[19] This important result was followed by the measurement of the contribution associated with the CNO cycle.[20]

While the controversy on the explanation of the solar neutrino problem was still ongoing, a major step forward in neutrino astronomy was made by the detection of the first extragalactic neutrinos (from the supernova explosion SN 1987A) observed by means of two water Čerenkov detectors (12 events in the electron energy range 6–36 MeV, by *Kamiokande* II, and 8 events in the electron energy range 20–40 MeV, by IMB).[21] Nowadays, especially through the

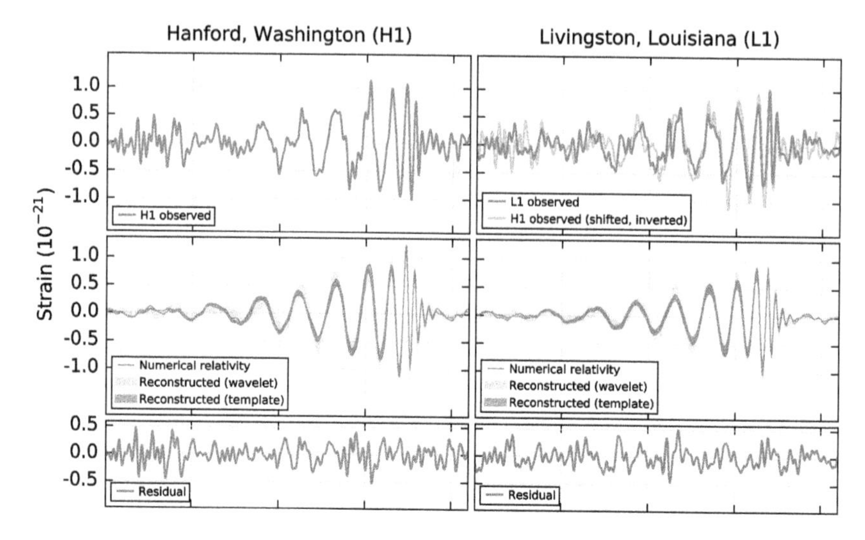

Figure 5.2 The first direct detection of gravitational waves; the horizontal coordinate represents time covering an interval of about 0.2 seconds (from: LIGO Scientific Collaboration and Virgo Collaboration. Abbott, B. P., Abbott, R., et al., "Observation of Gravitational Waves from a Binary Black Hole Merger," 2016. *Phys. Rev. Lett.*, **116**, id.061102; DOI: https://doi.org/10.1103/PhysRevLett.116.061102). A color version of this figure is available at www.cambridge.org/bertin

observations of detectors such as the IceCube Neutrino Observatory, completed in 2010 at the Amundsen–Scott South Pole Station, neutrino astronomy is in full bloom.

5.2 Direct Detection of Gravitational Waves

The first announcement of the positive detection of gravitational waves by the LIGO-*Virgo* collaboration, ascribed to the merger of two black holes of $\approx 30\ M_\odot$, was made in 2016 (Fig. 5.2) and referred to an event that occurred on September 14, 2015.[22] Other positive detections soon followed, but they lacked an electromagnetic counterpart, that is, detections of the event also by means of telescopes collecting electromagnetic radiation.

An extraordinary detection was made on August 17, 2017, and it is commonly referred to as GW170817. The announcement was given on October 16 of the same year. This is the first detection of a gravitational wave event by the LIGO-*Virgo* interferometer for which a clear counterpart was obtained by a number of observatories operating in various regions of the electromagnetic spectrum.

The gravitational wave signal lasted about 100 seconds, starting at 24 Hz and continuing up to a few hundred Hz. Combining the available information collected by *Virgo* (Italy), LIGO-Livingston (Louisiana, USA), and LIGO-Hanford (Washington State, USA) made it possible to localize the source in a region of the sky of approximately 28 square degrees.[23] The interpretation of the signal is that of the inspiral of two neutron stars with combined mass of ≈ 2.7–$2.8\ M_{\odot}$, at a distance of ≈ 40 Mpc; the final product of the merger may be either a black hole or a neutron star. The coalescence event was immediately associated with a Gamma Ray Burst of the short type (GRB 170817A) detected by the *Fermi* and INTEGRAL observatories, with a time delay of ≈ 1.7 s and with the S0 galaxy NGC 4993. The event clearly points to the long-term fate of systems such as the binary pulsar described in Section 3.3. Within a few hours, a bright optical transient was identified in NGC 4493 by several independent telescopes, with confirmed signals in the UV and in the near-infrared (NIR). Within a few days, X-ray (by *Chandra* and then XMM-*Newton*) and radio emission (by VLA) was detected at the location of the transient source. No related neutrinos were detected by means of the available facilities (in particular IceCube, ANTARES, and the *Pierre Auger* Observatory).[24]

About 1 year after,[25] it was noted that a previously recorded short duration Gamma Ray Burst event (GRB 150101B), located at a cosmological distance ($z = 0.1341$), presented significant analogies with the GRB associated with GW170817.

These findings suggest that these events may be related to what are called kilonovae, transient neutron star merger events with intermediate luminosity with respect to novae and supernovae.

5.3 The Solar Wind as a Collisionless Plasma

The subject of the solar wind deserves an entire book on its own.[26] At a distance from the Sun of approximately 1 AU, the plasma flow of the solar wind (primarily made of electrons, protons, and alpha particles) away from the Sun is relatively slow and irregular on the plane of the ecliptic (300–500 km s^{-1}), with temperatures at around 10^6 K, whereas outside the plane it is steadier, relatively fast (700–800 km s^{-1}), and slightly cooler. The particle density is of several particles per cubic centimeter. Typical values of the magnetic field intensity are of several nT (10^{-5} gauss). In the vicinity of the ecliptic an equatorial current sheet (corresponding to a true enhancement of the local plasma density) exhibits large-scale wavy behavior much like a "ballerina skirt," generating magnetic sectors, that is, regions of space where on the large-scale magnetic fields are characterized by opposite polarity.

Here, we wish to mention one special property that was soon brought out by *in situ* measurements performed by interplanetary missions, in particular by instruments on board the two Helios spacecrafts, operating in the period between the mid-1970s and the mid-1980s.[27]

The measurements clearly indicated that the velocity distributions of solar wind ions and electrons, as observed on the plane of the ecliptic at a distance from the Sun in the range of 0.3–1 AU, exhibited a variety of strong deviations from the canonical Maxwellian form and significant variability. In particular, the proton distribution was found to be often characterized by a velocity dispersion in the perpendicular direction (with respect to the direction of the interplanetary magnetic field) far larger than that in the parallel direction (i.e., a distinct temperature anisotropy) and by the presence of "double humps," that is, two coexisting streams, in the direction of the magnetic field. In turn, electrons often exhibited a distribution function with a cold core–hot halo structure and smaller levels of anisotropies. At higher energies, the distribution function of electrons often showed a distinct asymmetry (skewness) with excess velocities in the direction away from the Sun (sometimes called "strahl"). The general properties of the strahl were found to correlate with the overall structure of the solar wind and interplanetary magnetic field. The richness of the observed structures in phase space, with a mixture of coherent patterns and turbulent behavior, raises many specific questions about the physical mechanisms that originate the observed behavior and clearly points to phenomena that defy interpretations in terms of classical thermodynamically relaxed systems. In fact, it is easily realized that the ion meanfreepath for Coulomb collisions of the solar wind, even if it varies by some three orders of magnitude depending on the specific solar wind local conditions, may often be very large, on the order of 1 AU, so that it is no surprise that the solar wind exhibits many phenomena that reflect the dynamics of collisionless plasmas.

In many respects, such clear-cut evidence from *in situ* direct measurements shows that when the physical conditions demand it theoretical models may have to abandon simple thermodynamic arguments of the kind so often used in physics and astrophysics (e.g., see Section 1.4).

5.4 Gravity Assists and Other Curiosities

The general principle underlying the concept of the most standard gravity assists is very simple. An elastic gravitational encounter between two masses can be used to accelerate the lighter mass if we impose a suitable orbit, so that the lighter mass is deflected in the direction of the motion of the heavier mass. The gain in velocity has little to do with gravity and, instead, much to do with the simple process of an elastic bounce on a moving wall.

Consider a massive wall moving to the right, in the positive x direction, at speed V and an incoming ball hitting the wall, from the right, with velocity $u_{in} = -u$. If the ball is scattered back elastically, with velocity u_{out} in the positive x direction, it will keep the same kinetic energy (i.e., the same speed) in the reference frame in which the wall is at rest, $|u'_{in}| = -(u_{in} - V) = u + V = (u_{out} - V) = |u'_{out}|$, so that in the rest frame we have $u_{out} = u + 2V$.

If we now consider the inertial frame associated with the Sun and imagine an interplanetary probe moving with velocity $\vec{u}_{in}$ undergoing a gravitational encounter with a planet characterized by velocity $\vec{V}$, when the encounter is over (and thus the gravitational interaction term can be ignored) we expect $\vec{u}^2_{in} - 2\vec{u}_{in} \cdot \vec{V} = \vec{u}^2_{out} - 2\vec{u}_{out} \cdot \vec{V}$, so that the magnitude of $\vec{u}$, because of the encounter, is increased if the scalar product $2(\vec{u}_{out} - \vec{u}_{in}) \cdot \vec{V}$ is positive, that is, if the deflection occurs in the direction of the motion of the planet. Of course, this simple discussion is justified if the time of the encounter is very short with respect to the orbital periods of the planet and of the interplanetary probe.

The swing effect of gravity assists has been very often used in the navigation of many space probes, in particular of Mariner and Voyager missions, and, more recently, of various interplanetary spacecrafts (including *Cassini–Huygens*, *Ulysses*, and the *Parker* Solar Probe).

The encounter of a star with a binary star is a classical example of the richness of possible processes occurring in an unrestricted three-body gravitational problem. It can be shown that an encounter with a binary star characterized by low binding energy very often leads to the disruption of the binary, whereas the case of encounters with tightly bound pairs has a rather wide set of possible outcomes. As a general trend, tight pairs tend to become more tightly bound after an encounter, so that they often provide a gravity assist that accelerates an incoming star.[28]

In the 1980s it was noted[29] that a tightly bound binary star encountering in a Newtonian orbit a supermassive black hole (with mass comparable to that of the black hole that is thought to be hosted by the innermost regions of our Galaxy, $\approx 10^6 M_\odot$; see Chapter 7) may lead to a final state in which one of the two stars is captured into a bound orbit around the supermassive black hole whereas the other is ejected and accelerated to very high velocities, up to a few thousand kilometers per second. In this case, a gravity assist might occur at the center of our Galaxy to produce true hypervelocity stars, with a significant chance of eventually escaping from our stellar system. The phenomenon gained further appeal as a general phenomenon because, at the turn of the century, central supermassive black holes were found to be ubiquitous in galaxies. Yet it remained basically a piece of scientific curiosity until the year 2005, when the first hypervelocity star in our Galaxy was detected, a blue star with a heliocentric radial velocity of $\approx$ 850 km s^{-1}, and related to the ejection mechanism elaborated earlier.[30] Within a decade, dozens of hypervelocity stars had been detected[31] and by now, especially with the incoming data from the *Gaia* mission (see Section 2.3), the topic of hypervelocity stars has become one of the hot topics in modern astrophysics. Theory and observations are challenged to give evidence for or against proposed acceleration mechanisms. In addition, the detailed study of the observed orbit parameters in combination with the proposed acceleration mechanism can test quantitatively our current picture of the potential well (as provided by visible and dark matter) of the Milky Way Galaxy.

Notes

1 For example, see Perkins, D. 2003. *Particle Astrophysics*. Oxford University Press, Oxford; Grupen, C. 2005. *Astroparticle Physics*. Springer, Heidelberg.
2 Eddington, A. S. 1920. *Observatory*, **43**, 341; 1926. *The Internal Constitution of the Stars*. Cambridge University Press, Cambridge, UK. Other key references are Bethe, H. A. 1939. *Phys. Rev.*, **55**, 434; Chandrasekhar, S. 1939. *An Introduction to Stellar Structure*. University of Chicago Press, Chicago.
3 Fowler, R. H. 1926. *Mon. Not. Roy. Astron. Soc.*, **87**, 114.
4 Chandrasekhar, S. 1931. *Astrophys. J.*, **74**, 81; 1935. *Mon. Not. Roy. Astron. Soc.*, **95**, 207. See also Chandrasekhar, S. 1984. *Rev. Mod. Phys.*, **56**, 137.
5 An account of the pioneering work in this field can be found in the monograph written by one of the primary actors, Rossi, B. B. 1964. *Cosmic Rays*. McGraw-Hill, New York.
6 Fermi, E. 1949. *Phys. Rev.*, **75**, 1169; Bell, A. R. 1978. *Mon. Not. Roy. Astron. Soc.*, **182**, 147.
7 Greisen, K. 1966. *Phys. Rev. Lett.*, **16**, 748; Zatsepin, G. T., Kuz'min, V. A. 1966. *J. Exp. Theor. Phys. Lett.*, **4**, 78.
8 Bernabei, R., Belli, P., et al. 2018. *Nucl. Phys. At. Energy*, **19**, 307; Amaré, J., Cebrián, S., et al. 2019. *Phys. Rev. Lett.*, **123**, id.031301.
9 In the meantime, beautiful indirect evidence for the emission of gravitational waves was provided by the study of the binary pulsar briefly described in Section 3.3.
10 https://solarsystem.nasa.gov/missions/cassini/science/saturn/hexagon-in-motion/
11 Rossi, B., Olbert, S. 1970. *Introduction to the Physics of Space*. McGraw-Hill, New York.
12 Parker, E. 1958. *Astrophys. J.*, **128**, 664.
13 Starting with the investigation of the properties of the bow shock in the Earth's magnetosphere. See Bonetti, A., Bridge, H. S., et al. 1963. *J. Geophys. Res.*, **68**, 4017.
14 Bahcall, J. N. 1989. *Neutrino Astrophysics*. Cambridge University Press, Cambridge and New York. See also the interesting article by John Bahcall "How the sun shines," first published in 2000: www.nobelprize.org/prizes/themes/how-the-sun-shines-2
15 Curiously, in the standard solar model the luminosity, effective temperature, and radius of the Sun have been increasing monotonically in time, since the initial

ignition; in particular, in the last 1.1 Gyr, the solar luminosity has increased by $\approx 9\%$, the effective temperature by $\approx 0.5\%$, and the radius by $\approx 3\%$.
16 Nakahata, M. 1988. Ph.D. Thesis, *Search for ^8B solar neutrinos at Kamiokande II.* University of Tokyo.
17 Pontecorvo, B. 1968. *Sov. JETP*, **26**, 984; Gribov, V., Pontecorvo, B. 1969. *Phys. Lett.*, **B28**, 493.
18 Mikheyev, S. P., Smirnov, A. Yu. 1986. *Sov. J. Nucl. Phys.*, **42**, 913; 1986. *Il Nuovo Cimento*, **9C**, 17; 1986. *Sov. JETP*, **64**, 4. Wolfenstein, L. 1978. *Phys. Rev. D.*, **17**, 2369; 1979. *Phys. Rev. D.*, **20**, 2634.
19 The Borexino Collaboration. 2018. *Nature (London)*, **562**, 505.
20 The Borexino Collaboration. 2020. *Nature (London)*, **587**, 577.
21 Hirata, K., Kajita, T., et al. 1987. *Phys. Rev. Lett.*, **58**, 1490; Bionta, R. M., Blewitt, G., et al. 1987. *Phys. Rev. Lett.*, **58**, 1494; Hirata, K. S., Kajita, T., et al. 1988. *Phys. Rev. D*, **38**, 448; Bratton, C. B., Casper, D., et al. 1988. *Phys. Rev. D*, **37**, 3361.
22 LIGO Scientific Collaboration and Virgo Collaboration. Abbott, B. P., Abbott, R., et al. 2016. *Phys. Rev. Lett.*, **116**, id.061102.
23 Abbott, B. P., Abbott, R., et al. (LIGO Scientific Collaboration and Virgo Collaboration). 2017. *Phys. Rev. Lett.*, **119**, id.161101.
24 LIGO Scientific Collaboration and Virgo Collaboration and other Collaborations 2017. *Astrophys. J. Lett.*, **848**, id.L12.
25 Troja, E., Ryan, G., et al. 2018. *Nature Comm.*, **9**, id4089.
26 We may start with the classical reference to Hundhausen, A. J. 1972. *Coronal Expansion and Solar Wind.* Springer-Verlag, Berlin, Heidelberg, New York. See also Meyer-Vernet, N. 2007. *Basics of the Solar Wind.* Cambridge University Press, Cambridge, UK.
27 Marsch, E., Schwenn, R., et al. 1982. *J. Geophys. Res.*, **87**, 52; Marsch, E., Goldstein, H. 1983. *J. Geophys. Res.*, **88**, 9933. See also Marsch, E. 1984. *Plasma Astrophysics*, ESA SP-207, p. 33 (The European Space Agency, Paris); the second article in the book is an account on the discovery of the solar wind, by Rossi, B. 1984. *Plasma Astrophysics*, ESA SP-207, p. 27 (The European Space Agency, Paris).
28 Heggie, D. C. 1975. *Mon. Not. Roy. Astron. Soc.*, **173**, 729.
29 Hills, J. G. 1988. *Nature (London)*, **331**, 687.
30 Brown, W. R., Geller, M. J., et al. 2005. *Astrophys. J. Lett.*, **622**, L33.
31 Brown, W. R. 2015. *Annu. Rev. Astron. Astrophys.*, **53**, 15.

PART II

Dark Matter

6
Galaxies

The second part of this book is devoted to the study of one major problem in astrophysics that has challenged theorists and observers for more than a century. This is commonly known as the problem of dark matter. Currently, it is universally believed that dark matter plays an important role in determining the gravitational field from the scale of small galaxies up to that of clusters of galaxies and beyond. However, until we have established firmly what makes dark matter, we may argue that, in spite of the enormous successes achieved in astrophysics up to the level of cosmology, so far we still lack full understanding of the nature of gravity.

Gravity is a very weak interaction. It is a long-range force, with the same radial dependence as that of the electrostatic force; for two electrons, the ratio between the strength of their mutual gravitational and electrostatic interactions is $\approx 2.4 \times 10^{-43}$. Yet, on the grand scale of astronomical objects, gravity dominates the scene.

In view of our current understanding of the problem of dark matter, some pages written by Henri Poincaré and Richard Feynman on gravitation and on the dynamics of stellar systems are chillingly modern. In his essay *Science et Méthode* (1908), by referring to a mass estimate relevant to our stellar system, Poincaré notes (well before the discovery of galaxies) that "ce que ne donnerait la méthode de lord Kelvin, ce serait le nombre totale des étoiles, en y comprenant les étoiles obscures ...," thus anticipating that a possible discrepancy between observed and dynamical mass might be naturally ascribed to the presence of dark matter (see Section 7.1). In *The Feynman Lectures on Physics* (1963), Fig. 7-8 shows a picture of a globular cluster to illustrate that the law of gravitation is true at very large scales. Feynman simply states: "If one cannot see gravitation acting here, he has no soul." Then, after commenting briefly on the fact that, in spite of the apparent crowding of stars, "the distances between the centermost stars are very great and they very rarely collide," he continues to larger scales by means of Fig. 7-9, a picture of a spiral galaxy (M81). Here he adds: "The shape of this galaxy indicates an obvious tendency for its matter to agglomerate."[1] But he quickly qualifies his statements, by remarking: "Of course we cannot prove that the law here is precisely inverse square, only that there is still an attraction, at this enormous dimension, that holds the whole thing together." These pages were written well before the discovery of dark matter in galaxies (see Chapter 9). At the end of Chapter 10, we will highlight some ideas for an alternative to the existence of dark matter, at the cost of proposing a modification of Newtonian dynamics.

The discovery of dark matter should be assigned to the study of rotation curves of spiral galaxies. A posteriori, hints of the existence of dark halos come from a variety of observations well before the 1970s. In particular, we should mention two precursors of the problem of dark matter. Starting with the 1930s, the study of the solar neighborhood indicated the possible existence of some dark matter in the disk of our Galaxy[2] whereas the study of clusters of galaxies suggested the presence of "a great mass of internebular material."[3] As will be described in Chapter 8, evidence gathered from these precursors remained controversial until the end of the last century; in particular, for the problem of the solar neighborhood, our current perception is that the data do not show significant evidence for dark matter in the disk. Another study that contained clear clues about dark halos was the measurement, in the late 1950s, by means of 21-cm radio observations, of the rotation curve of M31, the Andromeda galaxy.[4] These data pointed to high velocities in the outer disk above the values expected from the visible mass present (see Fig. 3.1). Yet, the Schmidt model of our Galaxy,[5] which clearly lacked the presence of a dark halo, was the reference model used by dynamicists through the 1970s.

As will be described in Chapter 9, the systematic study of the rotation curves of spiral galaxies, especially with the advent of WSRT (*Westerbork Synthesis Radio Telescope*), eventually led to the discovery of dark halos. After first realizing that rotation curves tend to be flat in the outer parts and never show evidence of a Keplerian decline, a period of more than 10 years of uncertainty about the interpretation of the rotation curves veiled the discovery. In the end, HI observations extending well outside the optical disk were recognized to provide the decisive evidence that spiral galaxies are embedded in dark halos.

Therefore, this second part begins with a chapter that is intended to be a short introduction to galaxies. Galaxies "are to astronomy what atoms are to physics."[6] The chapter will start with a description of how galaxies were discovered, because the topic illustrates a sort of paradigm for more recent discoveries in astrophysics, such as that of GRBs (see Section 4.3). It will continue with highlights on their morphology, their structure and kinematics, following a description of their typical scales, with special reference to the concept of dynamical time scale. By now, all these should be considered classical information that is a prerequisite for any further discussion about the dynamics of galaxies. The subject of scales will give the opportunity to mention some modern developments about galaxies characterized by unusual properties that are currently studied at the frontier of galactic dynamics. The closing section of the chapter is devoted to introducing the virial theorem in its simplest form. This is an important tool at the the basis of many arguments about self-gravitating systems.

6.1 The Discovery of Galaxies and Their Morphological Classification

"What are galaxies? No one knew before 1900. Very few people knew in 1920. All astronomers knew after 1924."[7] The Shapley–Curtis debate about the nature of some nebulae, that is, extended sources of various appearance, such as M31, M51, M81, and M101, came to a clear-cut conclusion, in favor of Curtis, when Hubble convincingly determined the distance to some nebulae with the help of the study of Cepheids identified in them (see also Subsection 1.2.3; for an identification chart relative to the galaxy NGC 2403, see Fig. 6.3). Before such distance determination, various arguments had been put forward in favor of two alternative views. According to Shapley, the available observations and theoretical arguments suggested that the nebulae would be relatively small systems, in terms of size, luminosity, and mass. According to Curtis, the available observations pointed to the existence of huge and most luminous stellar systems, true "island-universes," that appear small and faint to our telescopes because of their enormous distance. For a while, the debate had remained undecided because of incomplete information about the nature of novae and supernovae and especially because of a claim, by Adrian van Maanen, of a direct measurement of rotation in M101, which eventually proved to be incorrect. Only in recent years, as a result of the fantastic progress in astrometry (see Section 2.3), can we actually detect the exceedingly small displacements (proper motions) that van Maanen had claimed to have observed. Note that at the distance of 1 Mpc a transverse velocity of 200 km s^{-1} produces in 1 year a displacement of $\approx 4.2 \times 10^{-5}$ arcsec = 42 μas.

6.1.1 Cepheids and Other Variables

The list of pulsating variables[8] comprises many different types of stars, among which are RR Lyrae (with period in the range from 1.5 to 24 hours) and

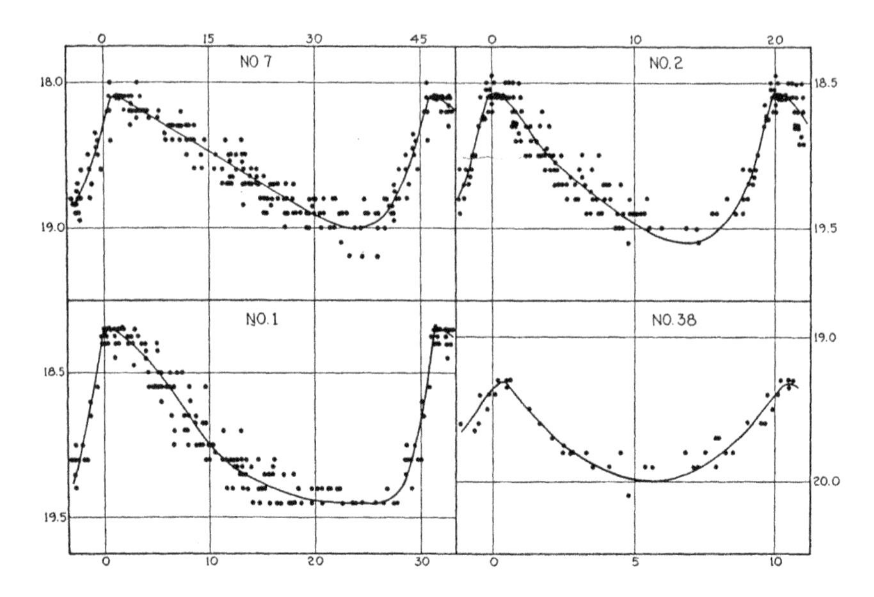

Figure 6.1 Light curves of four Cepheids in M31. Vertical scales represent photographic magnitudes; horizontal scales, days (from: Hubble, E. "A spiral nebula as a stellar system, Messier 31," 1929. *Astrophys. J.*, **69**, 103; reproduced by permission of the AAS).

Cepheids (with period in the range from 1 to 50 days). Astronomers distinguish between classical Cepheids (or Population I Cepheids),[9] of which notable examples are the stars δ Cep and Polaris, and other types, including Population II stars. At the time of the study by Hubble of Cepheids in nearby nebulae the distinction of the various classes was not so clear and this generated some incorrect conclusions in their use as standard candles.

Classical Cepheids are bright giants of spectral type F6 - K2. Their intrinsic luminosity ranges from 10^3 to more than $10^4\ L_\odot$, with amplitude of luminosity variation of up to 2 magnitudes, and so they are among the stars that are most easily identified. In order to have a feeling for the general character of the variability of these stars and the quality of the measurements in the present context, it is instructive to inspect the luminosity curves of four Cepheids in M31 shown by Hubble in his book *The Realm of the Nebulae*,[10] from which it is easily noted that brighter Cepheids are characterized by longer periods (Fig. 6.1).

6.1.2 The Period–Luminosity Scaling Law

Starting with the study of $\approx 2{,}000$ variables in the Magellanic Clouds by Henrietta Swan Leavitt at the beginning of last century, it had been realized that the period of Cepheids scales with their intrinsic luminosity, in the sense that more luminous Cepheids have longer periods. In order to apply the observed correlation as a standard candle, to determine the absolute luminosity of a Cepheid from a measurement of its period, astronomers need to calibrate the relation by studying some Cepheids at known distance, so that their absolute luminosity is known by independent means. This type of investigation was soon undertaken, beginning with the work of Hertzsprung in 1913. It has been an important line of research for about a century, especially because the use of Cepheids as standard candles finds application in modern observational cosmology: in particular, the related distance measurements lead to one determination of the value of the Hubble constant H_0.[11] In the context of the discovery of galaxies, an interesting account of the general issues involved can be found in the book *Galaxies*, by Shapley, in the revised version by Hodge (1971).

We wish to emphasize that, much like other scaling laws in astrophysics, the period–luminosity relation is interesting in its own right. Astronomers working on stellar structure and evolution should establish (and, by now, have well established) why and how certain stars such as Cepheids are characterized by this kind of variability and explain the origin of the observed scaling law. However, independently of its theoretical interpretation, the study of the scaling law requires attention to many important empirical aspects (including the dependence on the chemical composition of the pulsating star, the dependence of the relation on the wavebands considered, and the levels of extinction present along the line of sight in the observed objects) that are crucial for a correct application or, in different words, are a source of uncertainty in distance determination.

6.1.3 Distance Determination

At a meeting of the American Astronomical Society held from December 30, 1924, to January 1, 1925, it was announced that Hubble had identified Cepheids in M31 and M33, based on observations made with the 100-inch and 60-inch reflectors of the Mount Wilson Observatory. From the calibration of the period-luminosity scaling law available at that time, he derived a distance of ≈ 300 kpc. The modern estimate of the distance to M31 is ≈ 780 kpc ($\approx 2.5 \times 10^6$ light-years), that is, more than a factor of 2 larger than Hubble's estimate. In any case, this announcement gave the decisive evidence in favor of Curtis and marked the discovery of galaxies as island-universes.

6.1.4 Morphological Classification

In the 1920s, Hubble devised a morphological classification scheme for galaxies, often referred to as the tuning fork diagram, which is basically still in use today, in spite of many decades of investigations, improved observations, and several refinements. Given the time when it was conceived, the classification is based on optical (visible light) images. The diagram consists of a linear sequence of rather round objects, presumed to have a three-dimensional quasi-ellipsoidal shape, called elliptical galaxies, that are ordered from left to right according to the value of the observed apparent flattening (from E0 to E7, with the integer number defined as the number that best approximates 10 times the ellipticity $1 - b/a$, where a and b are the major and minor axes, respectively). From a transition class at the end of the En sequence, the so-called S0 galaxies, the two prongs of normal spirals (from Sa to Sc) and of barred spirals (from SBa to SBc) depart and continue to the right. The general change of morphology from left to right is often denoted as a change from early-type to late-type galaxies, although no special time-evolution character is currently associated with the terms used.

Ellipticals appear to lack special morphological features and do not show evidence for significant amounts of interstellar material and star formation (Fig. 6.2). In turn, spirals generally exhibit spiral structure, as well as the presence of interstellar gas, interstellar dust, and ongoing star formation (Fig. 6.3). In the original diagram by Hubble, the two spiral sequences are drawn in such a way that spiral structure is what today we would call a grand-design structure (with a pair of regular arms extending from the center to the periphery), with arms from tight to open as we move from category a to category c. In addition, either there is a central bar (and then we are on the SB prong) or the bar is absent (and then we are on the S prong). Spirals are characterized by the presence of a disk, which may or may not be accompanied by a central bulge (a sort of small central elliptical). The actual classification is more complex. First of all, grand-design spiral structure (which is dominant in galaxies such as M81 and M51) is not always present, especially in the visible; in many cases, such as in M31, M33, and M101, the observed spiral structure is less regular, and in others, such as NGC 2841 and NGC 7331, it is flocculent. Then the transition among the various categories occurs with continuity, so that a galaxy such as M83 may be called an intermediate type (modern notation would assign the symbol SA to normal spirals and SB to barred spirals, so that M83 might be called an SAB spiral). Finally, a variety of indicators are used to classify a spiral galaxy according to the $a \to c$ sequence that are not shown explicitly in the original tuning fork diagram.

The main indicators that define the $a \to c$ sequence are gas content (increasing), size of HII regions (increasing), arm spacing (increasing), size of nuclear bulge (decreasing), and visible mass (decreasing). If we take gas content and size of nuclear bulge as primary indicators, we find that the arm spacing

Figure 6.2 HST image of M60 (NGC 4649) and M60-UCD1 [credit: NASA, ESA and A. Seth (University of Utah, USA) "CC BY 4.0."]. The spiral galaxy NGC 4647 would partly show up at the upper-right of this image but is covered by the inset depicting M60-UCD1 in detail. M60 is the third-brightest giant elliptical galaxy of the Virgo cluster of galaxies and is classified as an E1-E2 galaxy. A color version of this figure is available at www.cambridge.org/bertin.

exhibits a triangular correlation with the $a \rightarrow c$ category.[12] In particular, a wide range of pitch angles characterizes Sc galaxies, from tightly wound spirals (such as NGC 5364; $i \approx 10$ degrees) to very open spirals (such as NGC 5247; $i \approx 30$ degrees), whereas only tight spirals are found as Sa galaxies.

Finally, as is clear from Part I, where we highlighted many aspects of the progress in astronomy associated with the access to different parts of the electromagnetic spectrum, the morphological classification introduced by Hubble refers to only normal galaxies, whereas modern astronomy includes the study of more exotic objects that were discovered in more recent times, such as radio-galaxies, quasars, and Active Galactic Nuclei, which are also galaxies or galaxy-related objects, but with characteristics that go well beyond those of the tuning fork diagram. In turn, normal galaxies, when studied outside the visible part of the electromagnetic spectrum, may also exhibit unexpected properties. For example, many bright ellipticals host a significant component of interstellar medium, present as a hot X-ray-emitting plasma.

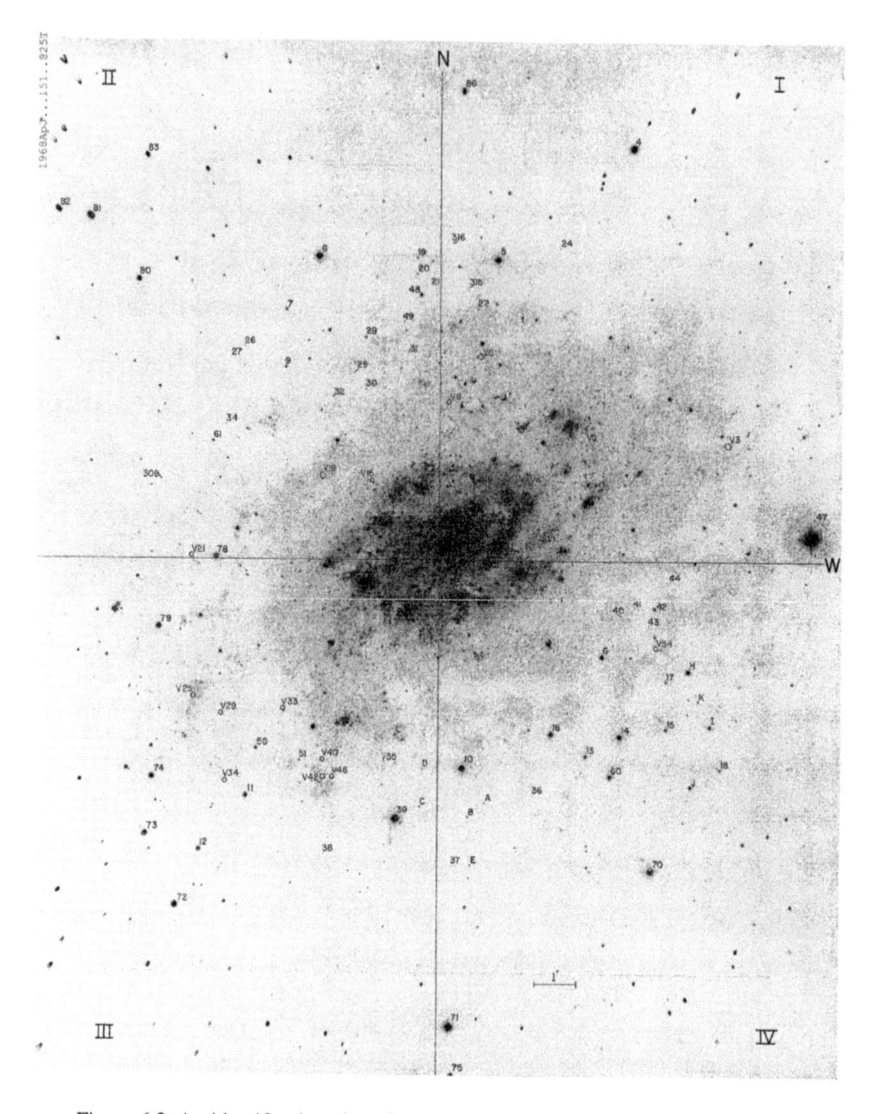

Figure 6.3 An identification chart for the spiral galaxy NGC 2403; 17 Cepheids are identified with their variable star numbers as reported in Table 5 of the source article (from: Tammann, G. A., Sandage, A. "The Stellar Content and Distance of the Galaxy NGC 2403 in the M81 Group," 1968. *Astrophys. J.*, **151**, 825; reproduced by permission of the AAS). NGC 2403 is an Sc spiral galaxy in the constellation Camelopardalis, at a distance of approximately 8 million light-years. As will be described in Chapter 9, the study of the rotation curve of NGC 2403 has played a key role in obtaining decisive evidence for the existence of dark halos in spiral galaxies.

6.2 Scales

Before summarizing the typical scales of galaxies, we should recall some obvious facts. Most of the information gathered by astronomers is from their light, which is dominated by the stellar component. Radio or X-ray emissions provide important information, but are generally associated with sources that can be considered a minority component in the overall mass budget. As will be demonstrated in Part II, we have convincing evidence that galaxies are embedded in massive halos that apparently extend to very large distances from the center, so that we are unable to determine their outer boundaries. Therefore, unless specified otherwise, the scales that we will mention as typical scales for galaxies will generally refer to the stellar component, that is, to the dominant visible component.

In a model-independent way we may define a linear size as the linear scale (radius) of the volume (or of the relevant isophote) that encloses half of the stellar component (see Subsection 1.2.3). This linear scale (sometimes defined as the half-light radius) is typically in the range of 1–20 kpc. There are some spiral galaxies, such as UGC 2885, for which we see starlight out to ≈ 100 kpc from the center, and dwarf galaxies, such as M60-UCD1, that are more compact than indicated by the scale range just given (inset in Fig. 6.2).

In terms of mass, the typical range is from 10^8 to $10^{12}\ M_\odot$. On the side of lower masses, there are stellar systems that comprise a variety of dwarfs and then the very regular globular clusters, generally found hosted by a normal galaxy. On the side of large self-gravitating systems, rather than individual galaxies we find galaxy groups, such as the Local Group, which hosts the Milky Way Galaxy, M31, M33, and more than 50 other galaxies (most of which are dwarfs); the Local Group extends on a scale of 1 megaparsec. Clusters of galaxies may comprise thousands of galaxies and thus be much more massive (up to $\approx 10^{15}\ M_\odot$).

Typical velocities (of a star in a galaxy or of the interstellar medium inside a galaxy) are in the range 50–500 km s^{-1}. Actually, the value of 500 is very rarely attained. It appears that higher velocities are observed in only exceptional circumstances. Note that typical velocities of galaxies inside clusters of galaxies are $\approx 10^3$ km s^{-1}. The fact that these velocities are much smaller than the speed of light suggests that, with the exception of specific phenomena generally on the very small scale of black holes and other compact objects, relativistic effects do not play a significant role in the evolution of galaxies.

Characteristic time scales for star orbits are in the range 10^7–10^9 yr. Because the concept of dynamical time and the related value of the relevant scales for galaxies play a special role in Part II, on this topic we will take a short digression in the next subsection. Here we only mention that for the grand scale of galaxy evolution other important time scales should be kept in mind. In particular, the age of a galaxy is generally identified with the age of its constituents and is rather close to the age of the universe, that is, the Hubble time (in terms of the Hubble constant, $1/H_0 \approx 10^{10}$ yr; see Section 2.2). For galaxies, the typical star–star relaxation time, that is, the time scale beyond which the cumulative effect of deflections in star orbits due to stellar encounters becomes significant, exceeds the Hubble time by orders of magnitude; therefore, in general galaxies are to be considered as collisionless stellar systems. In contrast, the HI component of the interstellar medium in spiral galaxies can be seen as being made of clouds of different sizes that collide inelastically with one another on a rather short time scale, smaller than the typical dynamical time. In this sense, the interstellar medium of spiral galaxies is clearly dissipative. Finally, in a list of relevant time scales we should mention the time scale for the evolution of single stars. Intrinsically bright stars tend to be characterized by short time scales; in particular, the bright O and B stars that mark the so-called Population I have a lifetime in the range of 10^6–10^7 yr, much shorter than the typical dynamical time.

What has been listed above, in this section, gives only a flavor of the scales that are involved when we study the properties of galaxies. In fact, because galaxies are complex systems, for certain phenomena we should be ready to consider scales that differ by orders of magnitude from those given above. This concept will be elaborated further in the section devoted to the structure and kinematics of spiral galaxies.

In order to complete this brief summary on the relevant galactic scales, we record some numbers for the Milky Way Galaxy. Our Galaxy and M31 (Andromeda) are the most massive galaxies of the Local Group, with mass exceeding $10^{11}\ M_\odot$. The distance from the Sun to the Galactic center is estimated to be $\approx$ 8 kpc, so that the Sun is approximately halfway from the center to the periphery of the outer stellar disk. The Sun and the nearby stars, the most easily visible in the night sky, are moving on quasi-circular orbits around the Galactic center, with a velocity $\approx$ 200 km s^{-1}, corresponding to a rotation period of $\approx$ 200 Myr. In the solar neighborhood, the Galaxy volume density is $\rho_\odot \approx 0.1\ M_\odot$ pc^{-3}, whereas the disk surface density is $\Sigma_\odot \approx 50\ M_\odot$ pc^{-2}, not far from the density of a normal sheet of paper. It has also been noted that, in terms of energy densities estimated in the solar neighborhood, the value of the kinetic energy density associated with rotation around the Galactic center exceeds by a factor of $\approx 10^3$ all other forms of energy densities, such as that associated with the Galactic magnetic field (such field is $\approx 5 \times 10^{-6}$ gauss). All these are meant to be order of magnitude estimates. The construction of a detailed model of our Galaxy is still in progress.

6.2.1 Dynamical Time

What is the precise definition of dynamical time, a quantity often mentioned in the context of the dynamics of galaxies? How can we define the dynamical time in a precise way, so that we may evaluate it with the help of suitable measurements? These questions are often asked, but go well beyond the normal use of the term, which is intended to identify only a characteristic time scale for star orbits. In other words, there is all the freedom to set the most convenient definition, once we refer to a given system and to a given problem.

If we take spiral galaxies, characterized by the presence of a thin disk, we actually have two separate time scales that can be identified and thus two separate natural definitions of dynamical time. Given the fact that the disk is in approximate rotation around its center, we may take a typical value of the rotation period of stars in quasi-circular orbits and call this the dynamical time; this is the dynamical time relative to the so-called equatorial plane (the disk). For example, for our Galaxy, because the Sun is neither close to the center nor close to the periphery of the stellar disk, we may take as dynamical time scale the time it takes to the Sun to make a turn around the Galaxy, that is, $\approx 2 \times 10^8$ yr. Of course, if we study a problem in the vicinity of the Galactic center the relevant dynamical time will be shorter, and, in contrast, if we consider questions related to the dynamics of the outer disk the dynamical time will be longer. In any case, because stellar orbits are typically quasi-circular, the period of rotation is a natural definition. However, we should recall that the geometry of the disk suggests the presence of a second dynamical time scale, because the disk is thin. In other words, there are two natural length scales (radius and thickness) and thus it is natural to refer to two different dynamical time scales. In addition to the equatorial period, just introduced, we may then define a vertical time scale associated with the oscillations of star orbits above and below the equatorial plane of the Galaxy. In a simple "slab model" of the disk (see Chapter 8), for a star like the Sun, which performs small vertical oscillations, remaining well inside the density distribution of the disk (which has a thickness of ≈ 1 kpc; see Chapter 8), it is very easy to prove that the frequency of the vertical oscillations (which are quasi-harmonic) is given by $\sqrt{4\pi G\rho_\odot}$, so that the related dynamical time is also naturally defined (about the meaning of the quantity $G\rho$ in the cosmological context, see also note 8 in Chapter 2). This time scale is shorter than that based on the rotation around the Galactic center.

For round stellar systems, such as elliptical galaxies or globular clusters, star orbits have a variety of shapes and only a few of them may be quasi-circular. For this reason, the dynamical time for these systems is often associated with the relevant crossing time. If V is a typical star speed and R is a typical scale for the round stellar system, then the dynamical time is naturally identified with the quantity $4R/V$. (This is also a natural definition in the study of clusters of galaxies, in which V is a typical velocity of a galaxy inside a cluster of radial scale R.) With the help of the virial theorem (to be introduced at the end of this chapter), it can be shown that this definition is consistent with that associated with the dynamical frequency $\sqrt{4\pi G\rho}$, where ρ is a typical density of the round stellar system.

The short discussion just given is mostly aimed at answering some natural questions about the definition of dynamical time for galaxies in the context of observations or of introductions to the subject. When dealing with the definitions adopted in numerical simulations of galaxies and other stellar systems we may wish to take other definitions of the dynamical time, which preserve its meaning but may be more specific and more convenient. For example, in the study of the dynamics of round stellar systems one definition of dynamical time often adopted is $GM^{5/2}/(-2E_{tot})^{3/2}$, where M is the total mass and E_{tot} is the total energy (kinetic plus gravitational) of the simulated stellar system.

6.2.2 Galaxies with Unusual Scales

Starting with the study of small galaxies in the neighborhood of our Galaxy and nearby galaxies, it was quickly realized that a variety of objects populate the low mass/size end of the galaxy distribution. *Dwarfs*, *dwarf spheroidals*, *compact dwarfs*, and *ultra-compact dwarfs* are the terms introduced to classify the variety of small galaxies that are observed. We will comment on these systems in Section 10.5.

Moving up to the scales of more normal galaxies, for spiral galaxies it was noted that they are characterized by a common value of central surface brightness.[13] Nowadays, these are called High Surface Brightness (HSB) galaxies. By the end of the century it was realized that there exists a family of Low Surface Brightness (LSB) galaxies, which has attracted the interest of observers and dynamicists.[14]

A curious class of low surface brightness galaxies with relatively large half-light radii (larger than 1.5 kpc), named Ultra Diffuse Galaxies (UDG; see Fig. 6.4), has been found to be associated with very low values of stellar luminosities, in spite of their relatively normal linear scales.[15] Curiously, when their kinematics has been examined,[16] very small velocities have been found, smaller than 40 km s^{-1}. The small velocities that are observed are indicative of lack of dark matter. This is a definitely interesting class of objects with unusual scales; however, their modeling and interpretation deserve deeper studies.

Finally, we wish to mention that on the high mass/size end of the galaxy distribution, a class of rapidly rotating spirals, called super spirals, has been

Figure 6.4 HST image of the Ultra Diffuse Galaxy NGC 1052-DF2 [credit: NASA, ESA, and P. van Dokkum (Yale University)]. A color version of this figure is available at www.cambridge.org/bertin

found at relatively large distances (redshift in the range 0.05–0.30).[17] These are galaxies that appear to be exceedingly large in linear scale, with rotation velocities obtained from optical spectra in the range 240–570 km s^{-1}. Especially because of their distance, also in this case modeling and interpretation await deeper investigations.

6.2.3 Galaxy Satellites

The presence of numerous companions of small mass is consistent with the modern picture of structure formation, even though the observed number of substructures tends to be too low with respect to the expectations of cosmological simulations.

For the Magellanic Clouds and nearest dwarf galaxies convincing evidence has been gathered for their being true satellites of the Milky Way Galaxy. One unexpected property that has first been noticed for small galaxies in the neighborhood of our Galaxy refers to their distribution. In fact, it had been argued that most of these satellites lie on a rather well-defined plane, much like in the solar system planets tend to lie on the plane of the ecliptic.[18] Such disks of satellites, confirmed by recent observations, have been shown by new investigations to occur also in nearby galaxies such as M31 and NGC 5128.[19] This is rather surprising and certainly unexpected in the standard cosmological model of structure formation.

6.3 Structure and Kinematics of Spiral Galaxies

If we examine pictures from *The Hubble Atlas of Galaxies* (176 galaxies, with reproductions from plates in the Mount Wilson–Palomar collection[20]) or more modern images taken by HST or major telescopes from the ground, we easily realize that galaxies are extremely complex systems.

By looking at optical pictures of galaxies such as M81, M51, M101, and M33, we recognize that spiral galaxies include stars of different types, HII regions of ionized gas, and dust, so that their structure is determined by the interplay of a stellar component and a complex InterStellar Medium (ISM). The most prominent structural features are a thin disk, a bulge (which may or may not be present), and spiral arms (which may or may not exhibit a grand design). The optical pictures are also suggestive of a significant ongoing star formation. With the exception of small spiral galaxies, the ISM, of which atomic hydrogen is the most massive component, contributes to only a fraction of the total visible mass, which is dominated by stars, especially evolved red giant stars.

Some general structural properties should be noted. As is well illustrated by edge-on galaxies, such as NGC 4565, NGC 5907, or NGC 891, and especially by bulgeless superthin galaxies,[21] disks are generally very thin and flat. Typically, the disk is made of an older and thicker component (Population II stars) and of a younger and thinner Population I. The stellar disk is often characterized by a thickness that is approximately constant at different radii. Atomic hydrogen and molecular gas generally define the thinnest disk component (although relatively recent observations have shown the existence of significant amounts of extra-planar atomic hydrogen); they are a clumpy medium, in which the most massive clumps (giant molecular clouds, with masses up to $10^5\ M_\odot$ and more) are thought to seed the regions of most active star formation. In many cases, atomic hydrogen is observed to continue as a regular flat disk well outside, beyond the bright optical stellar disk. From the geometrical point of view, prominent bars and spiral arms indicate that the disk is not axisymmetric. In many cases, the disk is clearly lopsided (e.g., see M101). In addition, especially in the outer gaseous disk, large-scale deviations from pure planar symmetry (often called galactic warps) are observed frequently.

In spite of such complexity, in most cases for spiral galaxies a useful and commonly applied simple model is that of a basic state that retains the general ingredients just mentioned but is conveniently characterized by axisymmetry (with respect to an axis passing through the galactic center, perpendicular to the plane of the disk), planar symmetry (symmetry by reflection with respect to the so-called equatorial plane of the disk), and stationarity (i.e., no evolution on the dynamical time scale). In terms of polar cylindrical coordinates, the gravitational field associated with such basic state can thus be derived from a mean potential $\Phi(R, z) = \Phi(R, -z)$, which is not known a priori, but is taken to be independent of time and of the azimuthal coordinate around the symmetry axis; the granularity of the gravitational field is ignored (especially the one associated with the discreteness of the stellar component; see the comment on the relaxation time scale at the beginning of Section 6.2).

Obviously, this does not mean that dynamicists should ignore time evolution and deviations from the assumed symmetry. As for any physical macroscopic system (the reader may refer to many well-known examples in fluid dynamics or plasma physics), the stationarity and the assumed symmetries of the basic state are only a convenient starting point for studying the real systems, where in many cases the relevant deviations can be treated as perturbations. In dynamical studies that address the possible existence of dark matter, it is especially important to focus on galaxies for which the viability of the simple basic state just described is best evident (also from a kinematical point of view; see further comments below in this section).

In this context, the surface brightness profiles of galaxy disks are approximately exponential:

$$I(R) = I_0 \exp(-R/h), \tag{6.1}$$

with two scale parameters, the central brightness I_0 and the exponential length h. The related total luminosity is then $L = 2\pi h^2 I_0$, half of which comes from the disk of half-light radius $R_e \approx 1.678h$. In standard astronomical units (magnitudes per seconds of arc squared), this empirical law becomes

$$\mu(R) = \mu_0 + 1.086\frac{R}{h}. \tag{6.2}$$

We commented earlier, in Section 6.2, on the role of μ_0 in separating the classes of HSB and LSB spirals. Deviations from a simple exponential profile are generally ascribed to the presence of a bulge (or other features, the discussion of which goes beyond the scope of this short introduction to the subject). The observed brightness is largely due to the stars, with some extinction by dust in the ISM. Different wavebands might then exhibit different profiles, in particular different exponential lengths. In practice, the small color gradients that are generally observed suggest that we may reasonably assign a mass-to-light ratio to the stellar component that is approximately constant with radius. In other words, the mass distribution of the stellar disk is generally thought to be characterized by an exponential surface density profile, proportional to $I(R)$. The contribution of the gas and of the bulge to the total mean gravitational potential have to be determined separately, because of their different spatial distribution.

Here we wish to re-emphasize a point that was briefly noted earlier, in Subsection 1.1.9. This is the importance of observations in the near-infrared (especially those in the K-band) in probing the structure of galaxies. Three main reasons justify this point: (1) in the near-infrared the images are least affected by dust extinction; (2) O, B, and other young stars if present dominate pictures taken in visible light, but contribute only little to the overall stellar mass, which instead is largely associated with evolved red giants and thus best traced by near-infrared observations; and (3) as we move further out to consider galaxies at cosmological distances, their stellar luminosity is more and more red-shifted.

In Section 4.4 we introduced some general concepts that form a useful framework for the description of the kinematics of spiral galaxies. In Chapter 9 we will describe in detail how studies of the kinematics of spiral galaxies have led to the discovery of dark matter halos. Here we only summarize some qualitative aspects that should help visualize an overall picture of the orbits that are associated with the motion of stars and gas in the mean potential Φ mentioned at the beginning of this section.

The disk of spiral galaxies is made of material (stars and gas) that is approximately performing a circular motion around the center of the galaxy, with small vertical oscillations around the equatorial plane of the disk. The term *approximately* is required not only because the assumed symmetry of galactic structure is only approximately realized in individual galaxies, but especially because stars and gas perform orbits that include deviations from average orbits (random motions). Qualitatively speaking, the disk is thin because it is cold, in the sense that the relevant random motions are typically much smaller than the typical average speeds. In Section 7.5 we will describe in some detail these deviations for star orbits. For the ISM, the flow is turbulent (nonlaminar) and thus also characterized by random motions that accompany the circular flow. These random motions are generally referred to as velocity dispersions and contribute to the pressure term present in the dynamical fluid equations. For atomic hydrogen and molecular gas the velocity dispersion is generally of a few kilometers per second; for the stars the values are higher and depend on the location in the disk (a typical velocity dispersion for the stars is 30 km s^{-1}).

6.4 The Virial Theorem for a Self-Gravitating System

The virial theorem is a general quantitative relation that specifies how various forms of energy present in a complex system participate in its dynamics. Frequently, it is considered under stationary or quasi-stationary conditions, so that it describes how the various forms of energy participate in the equilibrium of the system. It is generally derived by suitable summation (integration) of the relevant dynamical equations in the position (space) variables, in a variety of physical conditions. Here we will consider one of the simplest cases, that is, the virial theorem for a discrete system of N particles in mutual gravitational interaction (which is the relevant case for the study of stellar systems). In the closing subsection we will introduce the virial theorem for systems described as continuous media.

As will be illustrated by one important example in Chapter 8, the theorem allows us to make several arguments and to draw specific conclusions for many different astrophysical systems, independently of their detailed structure and microscopic dynamics. In particular, it allows us to estimate the mass of a self-gravitating system, starting from the measurement of its typical velocity and length scales. It also shows that, broadly speaking, self-gravitating systems behave as systems with negative specific heat.

6.4.1 N-body Systems

Consider a system of N point-mass stars in mutual gravitational interaction. Let $\vec{x}_i$, $\dot{\vec{x}}_i$, and $\ddot{\vec{x}}_i$ be the position, velocity, and acceleration vectors of the ith star. Furthermore, let $x_{ij} \equiv |\vec{x}_i - \vec{x}_j|$ be the distance between the ith and the jth star and $\vec{F}_{ij} = -Gm_i m_j (\vec{x}_i - \vec{x}_j)/x_{ij}^3$ be the gravitational force acting on the ith star by the jth star. Note that $\vec{F}_{ij} = -\vec{F}_{ji}$. As usual, a dot over a quantity represents derivative with respect to time.

The equations of the motion for the ith star can be written as

$$m_i \ddot{\vec{x}}_i = \sum_j^{j \neq i} \vec{F}_{ij}. \quad (6.3)$$

Here the indices i and j run from 1 to N; therefore, the above relation represents a set of $3N$ equations. If we take the scalar product by $\dot{\vec{x}}_i \cdot$ on each side of Eq. (6.3) and sum over the index i,

$$\sum_i m_i \dot{\vec{x}}_i \cdot \ddot{\vec{x}}_i = \sum_{i,j}^{j \neq i} \dot{\vec{x}}_i \cdot \vec{F}_{ij}, \quad (6.4)$$

we derive the energy conservation equation:

$$\frac{dK}{dt} \equiv \frac{d}{dt}\left(\sum_i \frac{1}{2} m_i \dot{\vec{x}}_i^2\right) = \frac{d}{dt}\left(\frac{1}{2}\sum_{i,j}^{j \neq i} \frac{Gm_i m_j}{x_{ij}}\right) \equiv -\frac{dW}{dt}. \quad (6.5)$$

The quantity K is the total kinetic energy, the quantity W is the total gravitational energy, and the conservation of energy is expressed by saying that $E_{tot} \equiv K + W$ is constant in time.

In the derivation of Eq. (6.5) from Eq. (6.4), we have made use of the definition of the gravitational force $\vec{F}_{ij}$ as a function of $(\vec{x}_i - \vec{x}_j)$, the relation $\vec{F}_{ij} = -\vec{F}_{ji}$, and the property

$$\frac{dx_{ij}}{dt} = \frac{d}{dt}\sqrt{(\vec{x}_i - \vec{x}_j)\cdot(\vec{x}_i - \vec{x}_j)} = \frac{1}{x_{ij}}(\vec{x}_i - \vec{x}_j)\cdot(\dot{\vec{x}}_i - \dot{\vec{x}}_j). \tag{6.6}$$

The virial theorem can be derived similarly, by multiplying (scalar product) Eq. (6.3) by $\vec{x}_i\cdot$ and by summing over the index i:

$$\sum_i m_i \vec{x}_i \cdot \ddot{\vec{x}}_i = \sum_{i,j}^{j\neq i} \vec{x}_i \cdot \vec{F}_{ij}. \tag{6.7}$$

The left-hand side can be rewritten in terms of the quantity

$$I \equiv \frac{1}{2}\sum_i m_i \vec{x}_i^2 \tag{6.8}$$

and of the total kinetic energy, so that Eq. (6.7) becomes

$$\frac{d^2 I}{dt^2} - 2K = \sum_{i,j}^{j\neq i} \vec{x}_i \cdot \vec{F}_{ij}. \tag{6.9}$$

Then we obtain

$$\frac{d^2 I}{dt^2} = 2K + \frac{1}{2}\sum_{i,j}^{j\neq i}(\vec{x}_i - \vec{x}_j)\cdot \vec{F}_{ij} = 2K + W. \tag{6.10}$$

Note that in the course of this derivation we have also proved that

$$W = -\frac{1}{2}\sum_{i,j}^{j\neq i}\frac{Gm_i m_j}{x_{ij}} = \sum_{i,j}^{j\neq i} \vec{x}_i \cdot \vec{F}_{ij}. \tag{6.11}$$

For the more general case in which the central force governing the interactions among the N particles is produced by a potential proportional to x_{ij}^{-s}, the resulting virial theorem would be written in the form

$$\frac{d^2 I}{dt^2} = 2K + sW^{(s)}, \tag{6.12}$$

where $W^{(s)}$ is defined by analogy with the self-gravitating case, keeping in mind the different power dependence of the interaction potential and the different interaction coupling constant. The self-gravitating case corresponds to $s = 1$ and coupling constant $-Gm_i m_j$.

In most cases, the virial theorem is applied under the hypothesis that the system is quasi-stationary, so that for a self-gravitating stellar system the theorem is identified with the condition

$$2K + W = 0. \tag{6.13}$$

Note that the total kinetic energy K scales as the total mass M of the system, whereas the total energy W scales as the square of the total mass M^2; in fact, if we multiply each stellar mass m_i by a scale factor λ, keeping all other quantities unchanged, we obtain $M \to \lambda M$ and from Eq. (6.5) we see that $K \to \lambda K$ and $W \to \lambda^2 W$. The simple remark just made shows why the virial theorem can be used to estimate the mass of a quasi-stationary self-gravitating system.

Furthermore, from the definition of total energy and from Eq. (6.13) we find

$$E_{tot} = -K = \frac{W}{2}. \tag{6.14}$$

Therefore, if we extract energy from a self-gravitating system, so that $E_{tot} \to E'_{tot} < E_{tot}$, the system tends to heat up, because $K \to K' > K$ and to contract, because $W \to W' < W$. This somehow counterintuitive behavior is often described by stating that self-gravitating systems are characterized by negative specific heat.

6.4.2 Continuous Mass Distributions

For systems made of very large numbers of particles, it is natural to refer to a continuous description, which may be either kinetic or fluid. For self-gravitating systems, we thus refer to the equations of stellar dynamics (for stellar systems) or to a fluid model (for stellar systems or for individual stars modeled as self-gravitating fluids).[22]

Without going through relatively tedious derivations that can be found in several books and basically follow the path outlined above for the discrete case, here we wish to recall some rather natural expressions that characterize the key quantities I, W, and K. We just note that the virial theorem for a simple one-component self-gravitating system, as expressed by Eq. (6.10), holds true also in the continuous description.

In terms of the mass density distribution $\rho(\vec{x})$, the obvious expression for the moment of inertia corresponding to Eq. (6.8) is

$$I \equiv \frac{1}{2} \int \rho(\vec{x}) \vec{x}^2 d^3 \vec{x}. \tag{6.15}$$

Then from the definition of gravitational potential

$$\Phi(\vec{x}) = -G \int \frac{\rho(\vec{x}')}{|\vec{x} - \vec{x}'|} d^3 \vec{x}' \tag{6.16}$$

the correct forms corresponding to the expressions [see Eq. (6.11)] for the total gravitational energy suitable for the continuous description are

$$W = \frac{1}{2} \int \rho(\vec{x})\Phi(\vec{x}) d^3\vec{x} = - \int \rho(\vec{x})\vec{x} \cdot \frac{\partial \Phi}{\partial \vec{x}} d^3\vec{x}, \quad (6.17)$$

where the symbol $\partial / \partial \vec{x}$ denotes the gradient with respect to the spatial coordinates.

Finally, for the concept of total kinetic energy we recall that we can distinguish between kinetic energy associated with mean (ordered) motions (K_{ord}) and kinetic energy associated with random (internal) motions (K_{int}), that is, $K = K_{ord} + K_{int}$. For the fluid model briefly introduced in Section 4.4, if we multiply by $\vec{x}\cdot$ the Euler equation and integrate over the spatial coordinates we can easily demonstrate that

$$K_{ord} = \frac{1}{2} \int \rho(\vec{x})\vec{u}(\vec{x})^2 d^3\vec{x}, \quad (6.18)$$

whereas the kinetic energy associated with random motions is provided by the pressure contribution

$$K_{int} = \frac{3}{2} \int p(\vec{x}) d^3\vec{x}, \quad (6.19)$$

so that for the continuous fluid case the virial relation corresponding to Eq. (6.10) becomes

$$\frac{d^2 I}{dt^2} = 2K_{ord} + 2K_{int} + W = 2K_{ord} + W + 3 \int p(\vec{x}) d^3\vec{x}. \quad (6.20)$$

As a global measure of the hotness of a stationary self-gravitating system we may thus refer to the quantity $K_{int}/|W|$ and note that the virial theorem imposes the condition

$$0 \leq \frac{K_{int}}{|W|} \leq \frac{1}{2}. \quad (6.21)$$

For nonsingular, spherically symmetric, continuous mass distributions, some simple relations can be easily derived and find frequent application. If a system is spherically symmetric with respect to the origin and we adopt standard spherical coordinates (r, θ, ϕ), we have $\rho = \rho(r)$ and $\Phi = \Phi(r)$. If the density distribution is known, we can calculate the mass contained in a sphere of radius r as

$$M(r) = 4\pi \int_0^r \rho(r') r'^2 dr', \quad (6.22)$$

so that the total mass is given by

$$M = 4\pi \int_0^\infty \rho(r') r'^2 dr'. \quad (6.23)$$

The moment of inertia follows from the definition given by Eq. (6.15):

$$I = 2\pi \int_0^\infty \rho(r) r^4 dr. \tag{6.24}$$

To calculate the gravitational energy W, it is convenient to refer to the second equality in Eq. (6.17) and recall that, from a simple application of the Gauss theorem,

$$\vec{x} \cdot \frac{\partial \Phi}{\partial \vec{x}} = r \frac{GM(r)}{r^2}. \tag{6.25}$$

Thus we find

$$W = -4\pi G \int_0^\infty M(r)\rho(r) r dr; \tag{6.26}$$

but from the first equality of Eq. (6.17) we can also write

$$W = 2\pi \int_0^\infty \Phi(r)\rho(r) r^2 dr. \tag{6.27}$$

If we take a homogeneous sphere of density $\rho(r) = \rho_0$ and radius R, from Eq. (6.23) and Eq. (6.22) we have, for $r \leq R$, $M(r) = (r/R)^3 M$ and, for $r \geq R$, $M(r) = M$. In addition, from Eq. (6.24) we have $I = (3/10)MR^2$ and, from Eq. (6.26), we obtain $W = -(3/5)(GM^2/R)$. Finally we note that the inner potential $\Phi(r)$ inside such a homogeneous sphere is harmonic

$$\Phi(r) = -\frac{GM}{R}\left(\frac{3}{2} - \frac{r^2}{2R^2}\right) = -2\pi G \rho_0 \left(R^2 - \frac{r^2}{3}\right). \tag{6.28}$$

A suitable constant makes it continuous with the potential associated with the empty space outside the sphere, where $\Phi^{(ext)}(r) = -GM/r$. If we insert expression (6.28) in Eq. (6.27) we double-check that $W = -(3/5)(GM^2/R)$.

One might wonder how diverse expressions such as Eq. (6.26) and Eq. (6.27) might be reconciled with each other. Therefore, it is interesting to note that, for regular mass distributions, the two expressions can be traced back to a third definition of total gravitational energy W, which identifies the gravitational energy density in terms of the gravitational field

$$W = -\frac{1}{8\pi G} \int \left(\frac{\partial \Phi}{\partial \vec{x}}\right)^2 d^3\vec{x}. \tag{6.29}$$

This is analogous to a well-known result in electrostatics, which follows in the general case from the first equality of Eq. (6.17) by means of Green's first identity and the use of the Poisson equation. In the spherically symmetric case, the result can be proved rather easily by means of integrations by parts and application of the Poisson equation (to relate the mass density to the gravitational field). For a self-gravitating sphere, we then have a third convenient expression for the gravitational energy

$$W = -\frac{G}{2}\int_0^\infty \frac{M(r)^2}{r^2}dr. \tag{6.30}$$

Again, for the homogeneous sphere we find

$$W = -\frac{GM^2}{2R}\left(\int_0^1 t^4 dt + \int_1^\infty \frac{1}{t^2}dt\right) = -\frac{3}{5}\frac{GM^2}{R}. \tag{6.31}$$

Curiously, whereas for the previous two expressions (6.26) and (6.27) the integrals that determine W depend only on the region of space where $\rho \neq 0$, for this last expression, in terms of the gravitational field, the determination of W depends on the contribution of the energy density also in the empty space outside the sphere.

Notes

1 Feynman also notes: "Incidentally, if you are looking for a good problem, the exact details of how the arms are formed and what determines the shapes of these galaxies has not been worked out." The reader interested in this topic may look up Bertin, G., Lin, C. C. 1996. *Spiral Structure in Galaxies: A Density Wave Theory.* MIT Press, Cambridge, MA. On this topic, more recent references are given in the very interesting article Peterken, T. G., Merrifield, M. R., et al. 2019. *Nature Astronomy*, **3**, 178, where the measurement of the pattern speed of the grand-design spiral structure in UGC 3825 is reported.

2 Oort, J. H. 1932. *Bull. Astron. Inst. Neth.*, **6** (238), 249.

3 In the words of Smith, S. 1936. *Astrophys. J.*, **83**, 23.

4 van de Hulst, H. C., Raimond, E., van Woerden, H. 1957. *Bull. Astron. Inst. Neth.*, **14**, 1.

5 Schmidt, M. 1965. In *Galactic Structure*, eds. A. Blaauw, M. Schmidt. University of Chicago Press, Chicago, p. 513.

6 These words are taken from the introduction of Sandage, A. 1961. *The Hubble Atlas of Galaxies.* Publication 618, Carnegie Institution of Washington, Washington, DC.

7 Sandage, A. 1961. op. cit.

8 See Cox, J. P. 1980. *Theory of Stellar Pulsation.* Princeton University Press, Princeton, NJ.

9 See Shapley, H. (revised by P. W. Hodge) 1971. *Galaxies.* Harvard University Press, Cambridge, MA.

10 Figure 8 in Hubble, E. 1936. *The Realm of the Nebulae.* Yale University Press, New Haven, CT. The figure is the same as Fig. 1 on p. 120 of the article by Hubble, E. 1929. *Astrophys. J.*, **69**, 103.

11 Benedict, G. F., McArthur, B. E., et al. 2007. *Astron. J.*, **133**, 1810. See also Freedman, W. L., Madore, B. F. 2010. *Annu. Rev. Astron. Astrophys.*, **48**, 673.

12 Kennicutt, R. C. 1981. *Astron. J.*, **86**, 1847.

13 Freeman, K. C. 1970. *Astrophys. J.*, **160**, 811. In the blue band, the value is $\mu_0 \approx 21.65$ mag arcsec^{-2} with a small scatter of ≈ 0.3 mag arcsec^{-2}.

14 van der Hulst, J. M., Skillman, E. D., et al. 1993. *Astron. J.*, **106**, 548; McGaugh, S. S., Bothun, G. D., Schombert, J. M. 1995. *Astron. J.*, **110**, 573; McGaugh, S. S. 1996. *Mon. Not. Roy. Astron. Soc.*, **280**, 337; de Jong, R. S. 1996. *Astron. Astrophys.*, **313**, 45; Tully, R. B., Verheijen, M. A. W. 1997. *Astrophys. J.*, **484**, 145.

15 See van Dokkum, P. G., Abraham, R., et al. 2015. *Astrophys. J. Lett.*, **798**, L45; Danieli, S., van Dokkum, P. 2019. *Astrophys. J.*, **875**, id.155. See also Sandage, A., Binggeli, B. 1984. *Astron. J.*, **89**, 919.

16 van Dokkum, P., Danieli, S., et al. 2018. *Nature (London)*, **555**, 629 measured the line-of-sight velocities of 10 luminous globular-cluster-like objects in the UDG galaxy NGC 1052-DF2 and derived a velocity dispersion of only $\approx$10 km s^{-1}; for a second UDG in the same group van Dokkum, P., Danieli, S., et al. 2019. *Astrophys. J. Lett.*, **874**, id.L5 measured an internal velocity dispersion (for stars and globular clusters) of only $\approx$ 5 km s^{-1}. Mancera Piña, P. E., Fraternali, F., et al. 2019. *Astrophys. J. Lett.*, **883**, id.L33 studied the HI kinematics of a sample of six UDG galaxies, located at a distance of 75–100 Mpc, and found circular velocities in the range 20–40 km s^{-1}. See also Mancera Piña, P. E., Fraternali, F., et al. 2022. *Mon. Not. Roy. Astron. Soc.*, **512**, 3230.

17 Ogle, P. M., Lanz, L., Nader, C., Helou, G. 2016. *Astrophys. J.*, **817**, id.109; Ogle, P. M., Jarrett, T., et al. 2019. *Astrophys. J. Lett.*, **884**, id.L11.

18 Lynden-Bell, D. 1976. *Mon. Not. Roy. Astron. Soc.*, **174**, 695.

19 For example, see Kroupa, P., Theis, C., Boily, C. M. 2005. *Astron. Astrophys.*, **431**, 517; McConnachie, A. W. 2012. *Astron. J.*, **144**, 4.

20 Obtained with the 60- or 100-inch telescopes on Mount Wilson or the 200-inch reflector or the 48-inch Schmidt telescope on Palomar Mountain. Sandage, A. 1961. op. cit.

21 Kautsch, S. J., Grebel, E. K., Barazza, F. D., Gallagher, J. S., III 2006. *Astron. Astrophys.*, **445**, 765.

22 Readers interested in these concepts may consult, among several books and review articles written on these topics, Ciotti, L. 2021. *Introduction to Stellar Dynamics*. Cambridge University Press, Cambridge, UK; Bertin, G. 2014. *Dynamics of Galaxies*, 2nd ed. Cambridge University Press, New York.

7
The Supermassive Black Hole at the Center of the Milky Way

The tracking of the orbit of a star around SgrA* performed in the course of about two decades is one of the most beautiful achievements of astronomy. It will be the focus of this chapter, for the following reasons. This measurement has led to the most convincing evidence for the existence of a supermassive black hole, at the center of our Galaxy, and to an accurate measurement of its mass. As such, this is one of the nicest discoveries of astrophysical dynamics. In addition, it is a simple and clear-cut example of the general dynamical paradigm that is used to determine whether a system contains some form of invisible matter coexisting with the visible matter. It also shows that, in some cases, the invisible matter that sometimes has been called missing mass may be in a form (a black hole) that has nothing to do with the concept of dark matter as is commonly envisaged in galaxies, clusters of galaxies, and cosmology. The problem of dark matter will be addressed in the next three chapters. Dark matter is thought to be an all-pervasive distributed component of the universe, so far not identified with any known specific constituent. In contrast, the missing mass required to explain the orbits of the stars in the neighborhood of the Galactic center is a very compact mass distribution, the characteristics of which match one well-defined type of object, which we call black hole.

After a short section on the dynamical paradigm that leads us to declare a discrepancy between mass present and visible mass, we will first recall some observations that for decades have suggested that our Galaxy should host a central supermassive black hole. We will then summarize the main characteristics of the more recent study of star orbits, and in particular of the orbit of a star called S2, close to the source SgrA*, and comment on the detection of central black holes in other galaxies. The final dynamical section will be devoted to some general concepts about orbits; it will also include a short description of quasi-circular star orbits in spherical or axisymmetric time-independent potentials.

7.1 The Dynamical Paradigm at the Basis of the "Discovery" of Invisible Mass

As will be exemplified in this and in the following two chapters, the discovery of invisible or dark matter generally results from a discrepancy noted between observed quantities with respect to the expectations associated with the use of the natural dynamical model suggested in a given astrophysical context. In most cases, the discrepancy can be summarized by saying that the velocities of some tracers (gas or stars) that are observed exceed those expected on the basis of the variety of objects that the telescopes are able to detect. To explain the discrepancy, we may try to propose alternative models. In the absence of reasonable solutions, we are left with two drastic options, that is, either to declare that some invisible mass is present in such a way to justify the observed dynamics or to declare that the laws of physics that we are using should be revised. Before reaching such a decisive stage, great attention should be paid to exploring the theoretical and the observational parts of the discussion because we are dealing with complex remote systems. The use of quotes accompanying the word *discovery* in the title of this section (the same remark would suggest the use of quotes also in the title of Chapter 9) is meant to underline a comment that was made earlier on several occasions (e.g., see the beginning of Chapter 6), that is, that a proper discovery should include a firm proof of what the invisible mass inferred from the observed discrepancy is made of.

In the example presented in this chapter, the fast tracers are stars in the vicinity of a well-studied radio source. The stars are running fast and in some cases are seen to reach velocities of more than 1,000 km s^{-1}. By a patient monitoring of the positions of some stars a simple model emerges as a natural framework for interpreting what is observed, that is, the model of elliptical Keplerian orbits. The model holds well, that is, it is capable of fitting the data in detail. However, it requires the existence of a supermassive compact object that is not observed in the region where the observations are performed, close to the center of our Galaxy. Eventually it is concluded that such invisible mass should be assigned to one supermassive black hole. The fit to the data by the model provides a measurement of the mass of the black hole. The case is interesting and instructive, but does not lead to the discovery of the presence of dark matter in the normal modern terminology.

In the first case studied in Chapter 8, the observations refer to clusters of galaxies and were first made in the 1930s. The dynamical model used was that of virial equilibrium (see Section 6.4). The galaxies for which it was possible to measure their line-of-sight velocity indicated motions inside the clusters too fast with respect to the expectations based on the cluster mass as could be estimated from the visible light of the galaxies observed in the clusters. The discrepancy suggested the existence of additional mass present in the clusters. Later, X-ray observations demonstrated the existence of great amounts of an IntraCluster Medium (see Section 4.2). Eventually, the application of a simple hydrostatic equilibrium model for the ICM and other indicators (such as gravitational lensing) led to the conclusion that the discrepancy between visible mass and dynamical model persists, so that clusters of galaxies are currently thought to host large amounts of dark matter.

In the second case studied in Chapter 8, the vertical equilibrium in the solar neighborhood is modeled in order to relate the observed thickness of the disk and the velocity distribution of the stars in the vertical direction to the self-gravity of the disk, that is, to the gravity associated with the density of matter in the solar neighborhood. The model is basically a variation on the hydrostatic model, which includes the presence of various components and stellar dynamics. For approximately 50 years, a discrepancy of a factor of two appeared to occur between visible mass and dynamically expected mass. Such discrepancy was interpreted as evidence for a dark component in the disk. Eventually, at the end of the century, new and accurate astrometric data from the *Hipparcos* mission (see Section 2.3) demonstrated that no significant discrepancy actually exists. This proved that the disk in the solar neighborhood does not contain significant amounts of dark matter.

The discovery of dark halos in spiral galaxies described in Chapter 9 is based on the study of rotation curves (see Section 3.1) modeled in terms of momentum balance in the radial direction, as described by Eqs. (4.10) and (4.11). In the outer regions, beyond the optical disk, the observed rotation curve remains too high with respect to the expectations based on the visible mass. It is concluded that spiral galaxies are embedded in dark halos, most likely characterized by a round and slowly declining density distribution ($\rho_{DM} \sim r^{-2}$).

7.2 Early Evidence for a Supermassive Black Hole at the Galactic Center

As noted at the beginning of Chapter 3, the first radioastronomical signals detected by Karl Jansky in 1932 were most likely associated with the powerful radio source now called SgrA*. That region of the sky, in the direction of the constellation of Sagittarius, was soon identified with the center of the Milky Way Galaxy.[1]

Many scientists were intrigued by the properties of the Galactic center and the possibility that a supermassive black hole would be present there. Some of them became attracted by this problem also in view of the progress in understanding the mechanisms of energy production in Active Galactic Nuclei (see Sections 3.4 and 4.1), with models developed in the late 1960s and early 1970s.[2]

Among the many important contributions to astronomy by Jan Oort, which range from the discovery of the differential rotation of our Galaxy to his studies of comets, a special role is played by his interest in the Galactic center and in galactic nuclei. From a number of observational clues (largely based on 21-cm radio observations of atomic hydrogen), he often speculated about the role of the central regions of spiral galaxies in shaping the large-scale structure observed in the disk.[3] He also realized that the center of our Galaxy likely hosts a very compact mass responsible for many observed phenomena observed in its vicinity and beyond. In his own words,[4] "An ultracompact radio source of radius smaller than ~10 AU is presumably the actual center of the Galaxy. It may have a mass in the order of five million solar masses, and there is a suspicion that it may contain the 'engine' responsible for the many expulsion phenomena observed throughout the central region." Eventually, Oort maintained that the evidence for a supermassive black hole was not compelling.[5] Charles Townes is best known for inventing the maser and for his work in quantum electronics. He also developed interest in and made important contributions to space spectroscopy and infrared astronomy in relation to the study of molecules in the interstellar medium, the occurrence of water masers in the astrophysical context, the measurement of the shape of nearby stars by special interferometric techniques, and the Galactic center.[6] In practice, the work carried out by Townes and his collaborators, especially by means of the study of the motion of ionized gas by means of the Ne II fine-structure line in emission at 12.8 μ, led to the correct conclusion that the center of our Galaxy hosts a supermassive black hole, with a mass close to the currently accepted value.[7]

7.3 The Orbit of the Star S2 around SgrA*

The discovery was made by monitoring more than 100 stars in the innermost arcsecond of the Galaxy with telescopes from the ground.[8] We recall that at the distance of the Galactic center 2.5″ correspond approximately to 0.1 pc. Because of the high extinction, up to 30 magnitudes, in the direction of these observations, the monitoring was best performed in the near-infrared (especially in the K-band; see comments in Subsection 1.1.9 and Section 6.3), where the extinction can be reduced to only 3 magnitudes. The extremely small size of the area of interest required the use of special astrometric tools and, in particular, the application of adaptive optics (see Subsection 1.1.7). The stars of this so-called S-cluster are characterized by a random distribution of the orientation of their orbital angular momentum vector. Curiously, most of them appear to be young early-type (B and O) stars, which raises an astrophysical problem on its own (the so-called paradox of youth). The continuation of Fig. 7.1 illustrates the general structure of the reconstructed orbits for a set of 20 stars of the S-cluster in a square of $1'' \times 1''$ centered on SgrA* considered at rest; among these, we can see the orbit of the star S2, which is further illustrated in the first frame of Fig. 7.1, contained in a rectangle of $\approx 0.1'' \times 0.2''$.[9]

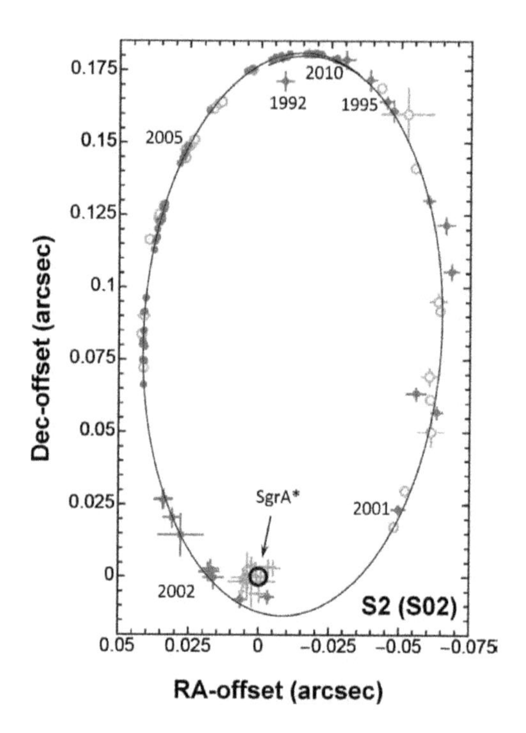

Figure 7.1 Above: The elliptical orbit of the star S2 in the vicinity of Sgr A* at the Galactic center (reprinted figure with permission from Genzel, R., Eisenhauer, F., Gillessen, S., "The galactic center massive black hole and nuclear star cluster," 2010. *Rev. Mod. Phys.*, **82**, 3121. © 2010 by the American Physical Society). The positions in the sky have been obtained in the period 1992–2010 by combining data from NTT/VLT with others from *Keck*. The pericenter of S2 is ≈ 125 AU (Astronomical Units), that is, ≈ 17 light-hours from a compact object with mass $\approx 4 \times 10^6\ M_\odot$. Continuation: A set of 20 star orbits in the vicinity of Sgr A* at the Galactic center (Gillessen, S., Eisenhauer, F., et al., "Monitoring stellar orbits around the massive black hole in the galactic center," 2009. *Astrophys. J.*, **692**, 1075; reproduced by permission of the AAS).

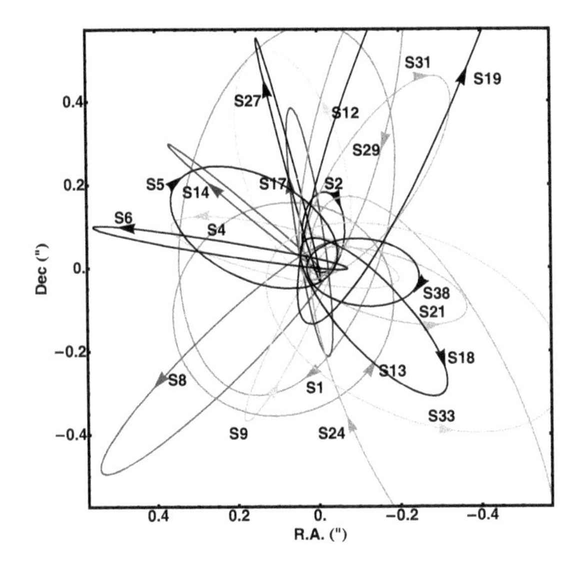

Figure 7.1 *(Continued)*

Let us focus on the orbit of the star S2 (a massive star of spectral type between B0 and B2.5, with estimated mass $\approx$ 15–20 $M_\odot$ and estimated radius $\approx$ 10 $R_\odot$). The orbital data-points are astrometric positions and velocities along the line of sight taken during a time interval of more than 15 years. With respect to SgrA*, the apocenter was reached around the year 1994 and again around the year 2010, implying an orbital period of approximately 16 years. Around the apocenter, the change in astrometric position is slowest. Such change is fastest around the pericenter, which was reached around the year 2002. To fit the data, a simple dynamical model is adopted, in which the orbit is modeled as a closed Keplerian ellipse. An order-of-magnitude estimate of the mass of the central object responsible for the gravitational field that determines such orbit is then obtained by applying Kepler's third law to the available information on the orbital period and on the semimajor axis, under a suitable assumption on the distance to the Galactic center R_0.

Several parameters can be adjusted to best fit the available data. The orbit is identified by six parameters, that is, the six conditions of the initial-value dynamical problem, which set the inclination of the orbit plane with respect to the line of sight, the shape and size of the ellipse, its orientation on the orbit plane, and the phase at which the orbit is performed. In addition, even in the simplest model of a potential well determined by a single point mass, the model requires the specification of the mass of the central object, its position, and its velocity.

More parameters would be involved for a description that includes the presence of an additional distributed mass and relativistic corrections to the gravitational field. A key parameter in the comparison of the observed data to the expectations of the dynamical model is the distance to the Galactic center R_0.

The fit to such a complex amount of data can be performed in different ways. Some difficulties remain with the interpretation of the S2 data obtained in 2002, close to pericenter, which also introduce uncertainties in the determination of R_0. Eventually, strong evidence emerges for the presence of a massive black hole (with mass $\approx 4 \times 10^6 M_\odot$), around which the star S2 performs an elliptical orbit in a plane inclined $\approx$ 45 degrees to the line of sight, reaching a speed of $\approx$ 7650 km s^{-1} at pericenter, $\approx$ 17 light-hours ($\approx$ 125 AU) from the compact object.[10]

7.4 Supermassive Black Holes in Other Galaxies

For a long time astrophysicists had argued that the powerful emission observed from quasars and Active Galactic Nuclei is associated with the presence of central supermassive black holes.[11] At the turn of the century, it became well established that central supermassive black holes are ubiquitous, even among quiescent, normal galaxies.[12] The case of the central black hole in the Milky Way Galaxy confirms this picture. The study of central black holes in external galaxies has been performed by a variety of methods, one of which makes use of maser emission produced in the vicinity of the black hole.[13] By now, this is becoming a general area at the frontier of astrophysical research. Below, we wish to comment briefly on only two particular exciting objects.

The presence of a supermassive central black hole at the center of the elliptical galaxy M87, in the Virgo cluster, has long been suspected[14] and eventually confirmed. This object has been the target of a major project that has led recently to the "image," by means of radio-interferometry, of the black hole on the scale of its Schwarzschild radius[15] of $\approx 3.6\ \mu$as, with a measurement of its mass as $\approx 6.5 \times 10^9 M_\odot$ at a distance of 16.8 Mpc (for the latest picture of the polarization pattern, see Fig. 7.2).

On a completely different scale, a very curious case is that of the ultra-compact dwarf galaxy called M60-UCD1, a presumed companion of the M60 elliptical galaxy also in the Virgo cluster (inset in Fig. 6.2). An apparently undisturbed spiral galaxy, NGC 4647, lies in the same general neighborhood. At a projected distance of 6.6 kpc from the center of M60, at an estimated distance of 16.5 Mpc from us, the half-light radius R_e (see Subsection 1.2.3 and Section 6.2) of UCD1 is only 24 pc ($\approx 0.6''$). With data obtained with the use

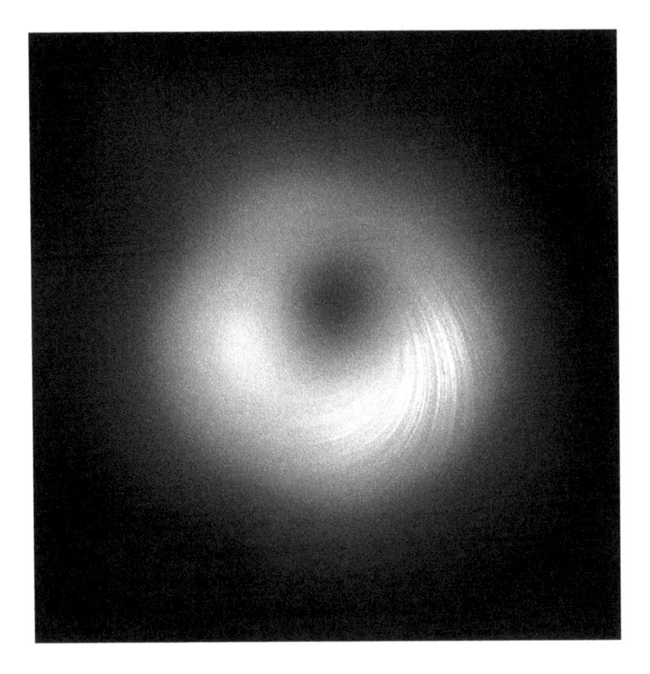

Figure 7.2 Polarized emission of the ring around the supermassive black hole at the center of M87. The lines mark the orientation of polarization, which is related to the magnetic field around the shadow of the black hole (credit: Event Horizon Telescope Collaboration). A color version of this figure is available at www.cambridge.org/bertin

of adaptive optics and detailed dynamical modeling, it is concluded that UCD1 hosts a central supermassive black hole of $\approx 2.1 \times 10^7\ M_\odot$, corresponding to $\approx 15\%$ of its total mass.[16] The galaxy M60 itself is thought to host a supermassive central black hole of $\approx 4.5 \times 10^9\ M_\odot$.[17]

7.5 Ellipses as Closed Orbits in Different Potentials

If we consider a time-independent central potential $\Phi(r)$, generated by a spherically symmetric distribution of matter $\rho(r)$, much like in the continuous description of a stellar system introduced in Subsection 6.4.2, the general nature of the orbit of an individual star [assumed to interact with the other parts of the stellar system only through the mean potential $\Phi(r)$] is easily assessed by a few simple considerations.

Given the symmetry of the potential, the specific total angular momentum of the star $\vec{J} \equiv \vec{x} \times \dot{\vec{x}}$ is constant in time, that is, it is an integral of the motion. Therefore, the relation

$$\vec{J} \cdot \vec{x} = 0 \tag{7.1}$$

is the equation of a plane with fixed orientation in space, called the plane of the orbit. The orbit can then be conveniently studied in polar coordinates (r, θ) defined in such plane, as a two-dimensional problem.

In the plane of the orbit, the study can be reduced to a one-dimensional problem. In fact, given the conservation of J, the specific energy $E = (v^2/2) + \Phi(r)$, also a conserved quantity, can be written as a function of only r and $\dot{r}$, in terms of the effective potential

$$\Phi_{eff}(r; J^2) \equiv \frac{J^2}{2r^2} + \Phi(r), \tag{7.2}$$

as

$$E = \frac{1}{2}\dot{r}^2 + \Phi_{eff}(r; J^2). \tag{7.3}$$

This is indeed a one-dimensional problem for the unknown $r(t)$, with the values of E and J^2 fixed by the adopted boundary (usually, initial) conditions on the equations of the motion.

For a regular density distribution characterized by finite mass the potential Φ is asymptotically Keplerian, that is, $\Phi(r) \sim -GM/r$, at large radii, whereas it is asymptotically harmonic at small radii. Normally, the density distribution is monotonically decreasing with radius and the related potential well is monotonically increasing with radius, with the minimum of the potential at the center, where the density has its maximum. In turn, for $J \neq 0$, the effective potential Φ_{eff} diverges at the origin, drops to a minimum value $E_0 < 0$ at a radius r_0, which for assigned Φ depends only on the value of J, and then, at large radii, matches the asymptotic Keplerian behavior of Φ.

Under the above-described conditions, for $E_0 < E < 0$, the orbit of a star is bound and contained in an annulus defined by two radii, r_{max} and r_{min}, defined as the radii where the radial velocity vanishes, that is, $\dot{r}^2 = 2(E - \Phi_{eff}) = 0$, called turning points of the motion.

Therefore, the radial motion is periodic (in general, not harmonic), characterized by radial period T_r

$$T_r = 2\int_{r_{min}}^{r_{max}} \frac{dr}{|\dot{r}|} = 2\int_{r_{min}}^{r_{max}} \frac{dr}{\sqrt{2(E - \Phi_{eff})}}. \tag{7.4}$$

The related radial frequency is then

$$\Omega_r = \frac{2\pi}{T_r}, \tag{7.5}$$

which, for assigned potential $\Phi(r)$, is a function of the specific energy and of the magnitude of the specific angular momentum, $\Omega_r = \Omega_r(E, J)$.

The angular frequency can be defined as the average value of $\dot{\theta}$

$$\Omega_\theta = \frac{2}{T_r} \int_{r_{min}}^{r_{max}} \frac{J}{r^2} \frac{dr}{|\dot{r}|}, \tag{7.6}$$

so that $\Omega_\theta = \Omega_\theta(E, J)$. In the inertial frame of reference orbits are closed if the ratio between the two frequencies is a rational number.

Some potentials are endowed with additional dynamical symmetries and generate simple closed orbits. It is well known that in a strictly Keplerian potential, $\Phi(r) = -GM/r$, bound orbits are closed and are ellipses with one of the two foci coinciding with the origin $r = 0$. For this potential one additional integral, the Lenz vector $\vec{L} = \dot{\vec{x}} \times \vec{J} + \Phi(r)\vec{x}$, reflects the constancy of the direction identified by the major axis of the ellipse, that is, the segment connecting the two points associated with the pericenter at $r = r_{min}$ and the apocenter at $r = r_{max}$.

Ellipses as closed orbits are not necessarily produced by a Keplerian potential. In fact, a central harmonic potential $\Phi = \Phi(r) = (1/2)\omega^2 r^2$ also produces closed orbits that are elliptical, characterized by frequency ω, but for these the origin $r = 0$ coincides with the center of the ellipse. In Chapter 6 we noted that the potential inside a homogeneous sphere is indeed harmonic [see Eq. (6.28)]. Therefore, orbits inside the core of a dark matter spherical halo (the term *core* for a density distribution generally describes a region where the density is approximately constant, ρ_0) are ellipses characterized by frequency $\omega = \sqrt{4\pi G\rho_0/3}$.

In conclusion, discovering that the orbit of a single star, such as S2, is a closed ellipse does not prove, strictly speaking, that the star is performing a Keplerian orbit around a compact mass. The proof comes from two main observed facts. Other stars appear to perform similar orbits in the same region and the centers of their orbits do not coincide with one another. More importantly, even if we did not have other star orbits to compare them with, from the observed velocities associated with the orbit of S2 it is clear that an ellipse produced by a harmonic potential is ruled out, because the star moves at different speeds (slow on one side, fast on the other) in the vicinity of the major axis.

Obviously, the general description provided in this section, which we started from the case of a spherically symmetric potential, also applies to star orbits on the equatorial plane of an axisymmetric disk associated with a mean potential

$\Phi(R, z)$; in terms of the standard three-dimensional polar cylindrical coordinates, on the equatorial plane ($z = 0$) the radial coordinate R can be identified with the radial coordinate r of polar coordinates in the plane.

7.5.1 Quasi-Circular Orbits

In general, the relevant potential $\Phi(r)$ is neither Keplerian nor harmonic. The solution for $r(t)$ can then be obtained in closed form by integration of Eq. (7.3), and the two-dimensional problem is solved by integrating the equation $\dot{\theta} = J/r(t)^2$. In practice, to draw the orbits we have to resort to numerical integration of these equations.

For the basic state that characterizes the disk of spiral galaxies we have a convenient tool to describe most of their stellar orbits in the plane of the disk without the need of numerical integration. In fact, the basic state is axisymmetric (see Section 6.3), and we can thus refer to the general framework just developed. In addition, the disk is dynamically cold (see Section 6.4), so that we can resort to the approximation of quasi-circular orbits, called the epicyclic approximation. This description is based on the fact that, statistically, most of the stars with given specific angular momentum J populate the bottom of the potential well defined by $\Phi_{eff}(r; J^2)$ (here, at variance with the notation of Section 6.3, we denote the radial distance from the center by r). Therefore, such a potential well is approximated by a parabola

$$\Phi_{eff}(r; J^2) \sim E_0 + \frac{1}{2}\kappa^2(r_0)(r - r_0)^2, \tag{7.7}$$

where

$$E_0 = \Phi_{eff}(r_0; J^2) = \frac{J^2}{2r_0^2} + \Phi(r_0). \tag{7.8}$$

If, for assigned $\Phi(r)$, we define the function $\Omega(r)$ by means of the relation

$$\Omega^2(r) \equiv \frac{1}{r}\frac{d\Phi}{dr}, \tag{7.9}$$

we note that for given J circular orbits are associated with the radius r_0 of the minimum of Φ_{eff}, so that

$$\Omega(r_0)r_0^2 = J \tag{7.10}$$

is an implicit relation for $r_0 = r_0(J)$ and confirms that Ω is the circular frequency associated with circular orbits. With this notation it is easy to prove that the quantity κ^2 appearing in Eq. (7.7) is given by

$$\kappa^2(r_0) = 4\Omega^2(r_0)\left[1 + \frac{r_0}{2\Omega(r_0)}\left(\frac{d\Omega}{dr}\right)_{r_0}\right]. \tag{7.11}$$

The frequency κ is called epicyclic frequency. Then the two frequencies $\Omega(r_0)$ and $\kappa(r_0)$ are implicit functions of J that correspond to the limit of $\Omega_\theta(E, J)$ and $\Omega_r(E, J)$, respectively, for $E \to E_0(J)$.

We may introduce the (assumed to be small) radial deviation r_1 from the circular orbit at r_0 as

$$r_1(t) = r(t) - r_0 \tag{7.12}$$

so that $\dot{r} = \dot{r}_1$, and the (also assumed to be small) angular velocity deviation $\dot{\theta}_1$ as

$$\dot{\theta}_1(t) = \dot{\theta}(t) - \Omega(r_0). \tag{7.13}$$

Then from Eq. (7.3) and Eq. (7.7) we see that r_1 performs harmonic oscillations of frequency $\kappa(r_0)$. At this point, it is easy to integrate the linearized equation for the conservation of angular momentum

$$r_0\dot{\theta}_1(t) = -2\Omega(r_0)r_1(t) \tag{7.14}$$

and show that the quasi-circular orbit can be seen as a composition of a circular orbit performed at frequency $\Omega(r_0)$ and a small elliptical epicycle, usually elongated along the direction of the motion (because the axis ratio $2\Omega/\kappa$ usually exceeds unity), run at frequency $\kappa(r_0)$.

In the two special cases discussed earlier in this section, we have $\kappa = \Omega$ for the Keplerian potential and $\kappa = 2\Omega$ for the harmonic potential. The factor 2 that makes the two cases different from each other marks the different nature of the corresponding closed elliptical orbits in relation to the location of the origin $r = 0$, as noted earlier in this section.

Notes

1 For example, see Oort, J. H., Rougoor, G. W. 1960. *Mon. Not. Roy. Astron. Soc.*, **121**, 171.

2 Among many papers on the subject, we wish to mention Ekers, R. D., Lynden-Bell, D. 1971. *Astrophys. Lett.*, **9**, 189.

3 Based on a study of NGC 4258 (M106), now known to host a central supermassive black hole, he argued that large-scale spiral arms in galaxies may be produced by material expelled at high speeds from its central regions; see van der Kruit, P. C., Oort, J. H., Mathewson, D. S. 1972. *Astron. Astrophys.*, **21**, 169. Oort credited Ambartsumian and Arp with being the first proposers of this picture.

4 Oort, J. H. 1977. *Annu. Rev. Astron. Astrophys.*, **15**, 295. In this review Oort refers to the recent results obtained by Wollman, E. R. 1976. Ph.D. Thesis, UC Berkeley; Wollman, E. R., Geballe, T. R., et al. 1976. *Astrophys. J. Lett.*, **205**, L5 and he adds: "These data permit a direct estimate of the mass inside ~ 0.4 pc, namely $\sim 5 \times 10^6 M_\odot$ ... If there is a black hole at the center this gives an indication of its mass. If there is no black hole it provides a check on the density distribution in Table 1."

5 Oort, J. H. 1985. In *IAU Symposium 106*, eds. H. van Woerden et al. Reidel, Dordrecht, The Netherlands, p. 349.
6 For the study of the Galactic center, see Wollman, E. R., Geballe, T. R., et al. 1977. *Astrophys. J. Lett.*, **218**, L103; Lacy, J. H., Baas, F., et al. 1979. *Astrophys. J. Lett.*, **227**, L17; Lacy, J. H., Townes, C. H., et al. 1980. *Astrophys. J.*, **241**, 132.
7 Genzel, R., Hollenbach, D., Townes, C. H. 1994. *Reports Progr. Phys.*, **57**, 417.
8 With observations started in the early 1990s. Among the most significant papers on the subject, we wish to mention: Schödel, R., Ott, T., et al. 2002. *Nature (London)*, **419**, 694; Ghez, A. M., et al. 2003. *Astrophys. J. Lett.*, **586**, L127; Ghez, A. M., et al. 2008. *Astrophys. J.*, **689**, 1044; Gillessen, S., Eisenhauer, F., et al. 2009. *Astrophys. J.*, **692**, 1075; Genzel, R., Eisenhauer, F., Gillessen, S. 2010. *Rev. Mod. Phys.*, **82**, 3121.
9 See also Fig. 13 in Gillessen, S., Eisenhauer, F., et al. 2009. *Astrophys. J.*, **692**, 1075.
10 Abuter, R., Amorim, A., et al. 2018. *Astron. Astrophys.*, **615**, id.L15. An interesting development in the monitoring of the S2 star has been the recent detection of the Schwarzschild precession associated with its orbit; see Abuter, R., Amorim, A., et al. 2020. *Astron. Astrophys.*, **636**, id.L5.
11 For example, see Frank, J., King, A., Raine, D. 2002. *Accretion Power in Astrophysics*, 3rd ed. Cambridge University Press, Cambridge, UK.
12 For example, see Ciotti, L. 2009. *Rivista del Nuovo Cimento*, **32**, 1.
13 For the case of NGC 4258 (M106), see Miyoshi, M., Moran, J., et al. 1995. *Nature (London)*, **373**, 127. Other interesting cases are those of NGC 1068 and NGC 3079.
14 See also the discussion of the nucleus of M87 by Sandage, A. 1961. *The Hubble Atlas of Galaxies*. Publication 618, Carnegie Institution of Washington, Washington, DC.
15 Akiyama, K., Alberdi, A., et al. (The Event Horizon Telescope Collaboration) 2019. *Astrophys. J. Lett.*, **875**, id.L6. For a study of the polarization of the ring, related to the morphology of the magnetic field configuration in the vicinity of the central black hole, see Akiyama, K., Algaba, J. C., et al. (The Event Horizon Telescope Collaboration) 2021. *Astrophys. J. Lett.*, **910**, id.L12 and id.L13.
16 Seth, A. C., van den Bosch, R., et al. 2014. *Nature (London)*, **513**, 398.
17 Shen, J., Gebhardt, K. 2010. *Astrophys. J.*, **711**, 484.

8
Two Precursors of the Problem of Dark Matter

This chapter describes two problems initially studied in the 1930s, a few years after the discovery of galaxies. In later years they were revisited and, a posteriori, appeared to be important precursors of the problem of dark matter, which in the meantime had emerged clearly from the study of the rotation curves of spiral galaxies (see Chapter 9). In reality, the early studies of clusters of galaxies and of the vertical dynamics of the solar neighborhood had failed to provide decisive evidence for the existence of dark matter. In fact, for both cases significant confusion remained until the end of last century. Then, with the advent of fresh data from the *Hipparcos* mission, it was proved that from the vertical dynamics of the solar neighborhood there is no evidence for the presence of dark matter in the disk. In turn, with the combined use of improved X-ray observations and of gravitational lensing, clusters of galaxies were eventually shown to be associated with large amounts of dark matter.

8.1 Clusters of Galaxies

The fact that the distribution of nebulae in the sky is often clumpy, suggesting the existence of systems of physically connected objects, had been recognized long before the discovery of galaxies.[1] After the discovery of galaxies, several studies addressed the properties of clusters of galaxies. Because initially the Hubble constant had been greatly overestimated (approximately by a factor of 8), the first estimates of the scales of these systems were basically all wrong,[2] yet the curiosity of the astronomers soon led them to raise important issues that even now, almost a century later, remain modern and to some extent open.

Clusters of galaxies host hundreds or thousands of galaxies within volumes characterized by linear scales on the order of 1 Mpc. In general, early-type (elliptical) galaxies are concentrated in their inner regions, whereas late-type (spiral) galaxies are found more frequently in their outer parts. In optical images clusters exhibit a frequently regular morphology, often with a round overall shape. They participate in the Hubble expansion, with an internal galaxy velocity dispersion on the order of 10^3 km s^{-1}. The study of clusters of galaxies soon became an area of research of special interest.[3]

8.1.1 Early Mass Estimates from the Virial Theorem

Here we wish to mention only briefly some of the points made in the pioneering work of the 1930s.[4] To measure the masses of galaxies and clusters of galaxies some bold but prophetic suggestions were made, in particular on the possible use of gravitational lensing as mass diagnostics (see Section 4 in the cited 1937 article by Zwicky "On the Masses of Nebulae and of Clusters of Nebulae") or the possible detection of a gravitational redshift associated with a cluster (see Section 5.4 in the cited 1933 article by Zwicky). In practice, the work followed two natural approaches, as is well described by Smith in the opening sentences of his 1936 article.[5]

To estimate the mass of a cluster of galaxies, we can start from the masses of the individual galaxies that belong to the cluster. By summing over the observed galaxies, we thus have an estimate of the total amount of the luminous (i.e., directly associated with the visible galaxies) mass present. In the 1930s, no direct evidence for additional mass was given by the available telescopes. Just after the discovery of galaxies, Hubble had provided estimates for the masses of galaxies, based on a sample of 400 extragalactic nebulae.[6] In simpler terms, if we assume, as we now know, that the visible mass of galaxies is dominated by the stellar component, we can estimate the total mass associated with individual galaxies by multiplying their observed optical luminosity by a mass-to-light ratio that, in solar units, is of order unity. Of course, to perform this estimate, we must know the distance to the cluster, so as to be able to convert the observed apparent luminosities into absolute luminosities. Various factors may be taken into account in view of more accurate estimates, in particular the possibility of intervening absorption/extinction affecting the apparent luminosities and the possibility of different mass-to-light ratios for different galaxies. In the spirit of the discussion of Section 7.1 we may call the mass thus estimated the observed mass of the cluster.

In the second approach, for many galaxies belonging to a cluster we may collect line-of-sight velocity data $\{v_{los}^{(i)}\}$ $(i = 1, 2, ..., Q)$ and angle positions in the sky $\{\vec{\theta}^{(j)}\}$ $(j = 1, 2, ..., P)$. The velocity measurements are difficult and time consuming, whereas position measurements are relatively easy, so that typically $Q \ll P$. Then we proceed by applying the dynamical constraint of the virial theorem, as described in Section 6.4.[7] To do so, we first evaluate the mean line-of-sight (recession) velocity of the cluster

$$u_{los} = \frac{1}{Q}\sum_{i=1}^{Q} v_{los}^{(i)} \tag{8.1}$$

and the angle position of what we may call the center of the cluster

$$\vec{\theta}_{cl} = \frac{1}{P}\sum_{j=1}^{P} \vec{\theta}^{(j)}. \tag{8.2}$$

The related velocity dispersion can be defined as

$$\sigma_{los}^2 = \frac{1}{Q}\sum_{i=1}^{Q} (v_{los}^{(i)} - u_{los})^2. \tag{8.3}$$

Similarly, from the angle position distribution we can compute an angle position dispersion; for a given distance to the cluster, we can convert this quantity into a projected radial scale of the cluster R_{cl}. Under the assumption of quasi-stationarity, the virial constraint can then be written as

$$GM_{cl} = k_{vir}\sigma_{los}^2 R_{cl} \tag{8.4}$$

and provides the desired dynamical mass estimator. Here the quantity k_{vir} is a dimensionless coefficient of order unity, the value of which depends on the internal structure and orbital composition of the cluster. Such virial coefficient can be further constrained by additional modeling that can be tested to some extent by the available observations.

In the 1930s it was immediately noted that a large discrepancy existed between the value of the dynamical mass M_{cl}, obtained from dynamical arguments such as the virial theorem, and the observed mass, obtained from the first approach. For the Coma cluster Zwicky noted a discrepancy by a factor of 400, whereas for the Virgo cluster Smith found a discrepancy by a factor of 200. Both scientists conclude by stating that the observed discrepancy might be interpreted as due to the presence of nonluminous or dark matter. Either such dark matter is to be associated with the individual galaxies (in other words, Hubble's arguments lead to a severe underestimate of galaxy masses) or it is possible "that the difference represents a great mass of internebular material within the cluster."

Initially the statistics were very poor. For Coma, as recorded by Zwicky (1933) and by Hubble and Humason (1931), the line-of-sight velocity was available for only four galaxies. Smith's dynamical analysis of the Virgo cluster comprised only 32 kinematical data-points. Yet, the mass discrepancy was clearly noted. Furthermore, in spite of the poor statistics, Smith addressed other important issues. In particular, from the fact that he could not detect a significant difference in dispersion between brighter and fainter cluster members he concluded that the Virgo cluster is likely not to be in a state of energy equipartition. From the distribution of velocities in the sky, he also concluded that the Virgo cluster does not show evidence for rotation or other systematic motions.[8]

Nowadays, modern estimates based on the virial theorem would not only include much higher numbers of kinematical data-points, but also a more thorough discussion of the modeling of the density distribution and kinematics of galaxies inside the cluster and especially of the problem of identifying real cluster members (as opposed to interlopers possibly present in the observed field in projection, but not belonging to the cluster). However, as we briefly saw in Section 4.2, in the 1970s a major surprise changed the picture altogether.

8.1.2 Modern Estimates of the Mass of the IntraCluster Medium

The discovery by X-ray telescopes of the IntraCluster Medium showed that indeed large amounts of internebular material are present in clusters of galaxies. Great attention was given to the emission processes (the key emission mechanism being recognized to be the free-free emission from a hot intergalactic plasma) and to the optical properties of the emitted radiation (optically thin radiation through the ICM), so that from the observed X-ray intensity maps it was possible to estimate the amount of mass associated with the ICM. With respect to the simple hydrostatic equilibrium condition, more advanced models were devised to describe the possibility of an internal cooling flow developing in the dense innermost regions of the clusters, where the cooling time associated with the emitted radiation may become shorter than the relevant dynamical time.[9]

In practice, the basic picture of quasi-hydrostatic equilibrium appeared to provide a reasonable framework for determining the total mass distribution inside clusters, following the steps outlined in Subsection 4.4.1. It was soon realized that, in the overall mass budget inside a cluster, the mass associated with the ICM exceeds the mass associated with the galaxies. Initially, the difficulties in determining the pressure gradients for the X-ray-emitting plasma (especially, the difficulties associated with the measurement of temperature and temperature gradients) were underestimated. For the most luminous X-ray cluster, RX J1347.5-1145, at redshift $z = 0.451$, it was found that "at 1 Mpc the ratio of gas-to-total-mass is 34%."[10]

In the meantime, at the turn of the century, the total masses of clusters of galaxies became the target of weak-lensing investigations (see Subsection 3.4.1). Initially, a trend was noted, indicating that total mass estimates based on weak lensing were systematically (slightly) larger than those based on X-ray analyses; the discrepancy was generally ascribed to the fact that weak lensing is likely to be sensitive to mass projected along the line of sight not directly associated with the clusters.[11]

Eventually, X-ray and gravitational lensing mass measurements converged and the discussion of the mass distribution in clusters of galaxies settled down into a generally accepted picture according to which (i) the mass in the form of IntraCluster Medium exceeds that in the form of galaxies and (ii) the amount of visible mass (i.e., in a first approximation the mass of the ICM) makes only 15% of the total cluster mass. In other words, $\approx 85\%$ of the total mass is believed to be in the form of dark matter.

8.1.3 The Relative Concentration of Dark and Visible Matter

In Chapter 9 we will see that the halos of dark matter in galaxies are thought to be diffuse, in the sense that the contribution of dark matter to the gravitational field that determines the observed rotation curves is thought to become more and more important as we move toward larger and larger radii. The issue that we wish to mention here is whether a qualitatively similar description can be given for the relative distribution of dark and visible matter in clusters of galaxies.

On the small scale of the central ≈ 100 kpc this general issue can be addressed in a variety of ways, especially by means of the use of strong lensing (for distant clusters; see Subsection 3.4.1), but we leave this topic (often referred to as the issue of the central cusp) to more advanced introductions to the problem of dark matter; furthermore, on such a small scale the ICM is likely to contribute less to the visible mass than the galactic mass.

The general large-scale distributions of the ICM plasma density and of the required dark matter density have been proved to admit a very simple test, which can be seen as a sort of exercise on the model of hydrostatic equilibrium. The test has been performed on two well-studied clusters, A496 and the Coma cluster, which can be taken as representative of the two categories of cool-core and non-cool-core clusters.[12] In a first simplified model galaxies are ignored and both the ICM and the dark matter density distributions are modeled with a simple expression that approximates the distribution of a regular isothermal sphere

$$\rho_i = \rho_i^{(0)} \frac{1}{[1 + (r/r_i)^2]}, \tag{8.5}$$

where subscript i takes on the values $i = ICM$ for the X-ray emitting plasma and $i = DM$ for the dark matter component. The ratio between the two length scales r_i found in the fit determines whether dark matter is more diffuse or

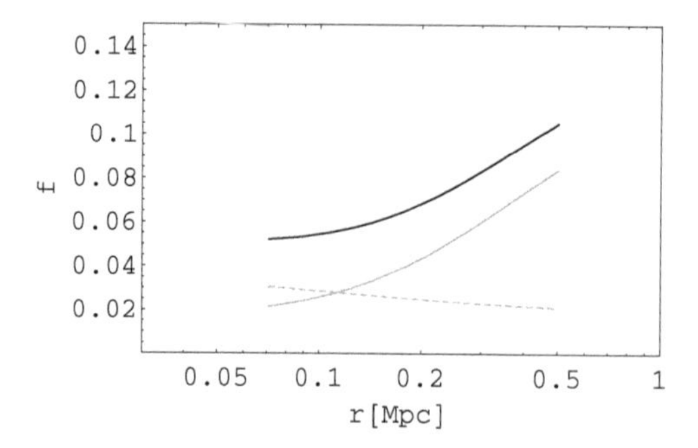

Figure 8.1 Galaxy fraction (thin dashed line), gas fraction (thin solid line), and visible mass fraction (thick solid line) for a simple model of the Coma cluster (figure taken from De Boni, C., Bertin, G., "The relative concentration of visible and dark matter in clusters of galaxies," 2008. *Il Nuovo Cimento B*, **123**, 31, DOI 10.1393/ncb/i2008-10506-x. © Società Italiana di Fisica, 2008; with kind permission). The figure is based on a Hubble constant corresponding to $h = 0.7$.

more concentrated than the visible matter. The test starts from fitting available estimates of the electron number density profiles in the two clusters to set the scales of the density distribution for the ICM. Then, for the assumed form of the distribution of dark matter, it applies the condition of hydrostatic equilibrium to predict the temperature profiles and thus to fit the available temperature points estimated from the X-ray spectroscopic data.[13] This study shows that, under such very simple modeling framework, the fit to the available data is very good (comparable to fits that are obtained by assuming for the distribution of dark matter profiles that are currently preferred in the cosmological context; see Subsection 10.3.1) and the dark matter distribution is more concentrated than that of the visible matter in both clusters. In particular, for the Coma cluster (for a Hubble constant of 70 km s^{-1} Mpc^{-1}) the fraction of ICM mass with respect to the total mass increases monotonically with radius, from 0.022 at 71 kpc to 0.084 at 500 kpc in the radial range where the test has been performed (Fig. 8.1). "The conclusion that dark matter is more concentrated than visible matter remains valid even when the contribution of the galactic component to the visible mass is considered explicitly."

Currently, the use of excellent new multiwavelength photometric and spectroscopic data and of much more sophisticated modeling techniques (including gravitational lensing) allows astronomers to disentangle the different roles of the different components of individual galaxy clusters with extraordinary precision.[14] Interestingly, for one of the best studied clusters, the MACS

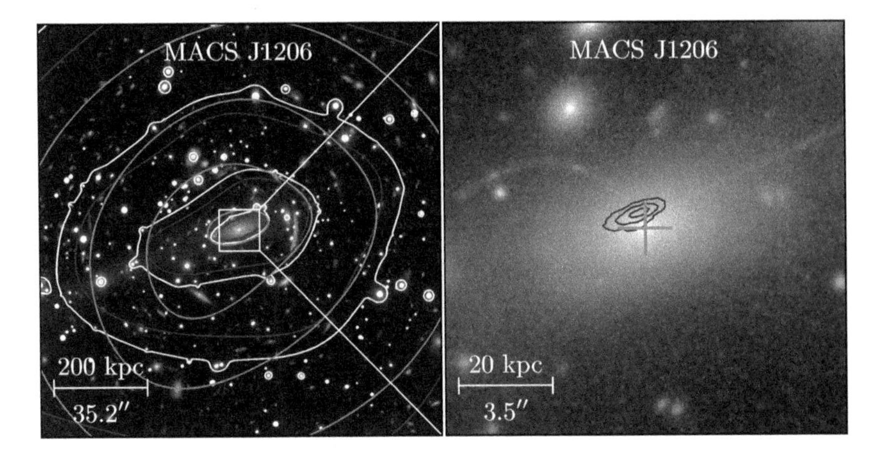

Figure 8.2 The central region of the cluster MACS J1206.2-0847. The various contours describe the properties of the surface mass density for total, diffuse dark matter, and ICM superposed to an HST WCF3+ACS image of the cluster. The right-hand frame is a zoom-in view. The plus sign indicates the position of the center of the brightest cluster galaxy. In the right-hand frame the contours show the 1σ, 2σ, and 3σ confidence levels of the diffuse dark matter component density peak. A full description is given in the caption to the color version of the figure (Bonamigo, M., Grillo, C., et al., "Dissection of the collisional and collisionless mass components in a mini sample of CLASH and HFF massive galaxy clusters at $z \approx 0.4$," 2018. *Astrophys. J.*, **864**, id.98; reproduced by permission of the AAS) from which the present figure is taken.

J1206.2-0847 cluster, the peaks of the density distribution of the three main components (dark matter, ICM, and galaxies) are found to basically coincide with the center of the central brightest cluster galaxy (Fig. 8.2), at variance with the case of the Bullet cluster illustrated in Fig. 4.2 (on which we will comment in Section 10.7).

8.2 The Thickness of the Galactic Disk in the Solar Neighborhood

Soon after it was realized that the Milky Way represents an edge-on view, from the inside, of the disk of our Galaxy, a program was started by Jan Oort to test quantitatively, in the solar neighborhood, the relation between the thickness of the Galaxy and the observed distribution of vertical stellar motions.[15] As anticipated in earlier investigations,[16] such a program would then lead to a dynamical determination of the mass present in the vicinity of the Sun.

A very simple toy model may help visualize the physical basis of the argument. If we consider an infinitely thin sheet of mass density $\Sigma_\odot$ as a model of the galactic disk, in its vicinity (much like for a capacitor) the vertical gravitational acceleration exerted by the disk can be approximated by $g = 2\pi G\Sigma_\odot$. Then stars moving with vertical speed v_z would be expected to reach a distance $h = v_z^2/(2g)$ from the thin sheet. Therefore, by treating in a similar way a large set of stars, we may argue that the finite thickness associated with their vertical distribution is directly proportional to the square of the vertical velocity dispersion and inversely proportional to the surface mass density of the disk. This argument is basically correct, but ignores some important aspects of the actual state of the disk, in particular the role of self-consistency: the above model computes the gravity acting on the stars from an infinitely thin sheet model representing the disk, whereas the disk has a finite thickness, associated with the star motions, which makes the thin sheet model quantitatively inadequate. In the next section we will illustrate the role of self-consistency by means of a well-known dynamical model.

Starting with the 1930s, the study of the vertical distribution and velocity dispersion for selected classes of stars in the solar neighborhood was then used, within a complex and detailed model, to derive a dynamical estimate of the mass density in the vicinity of the Sun. Oort noted a significant discrepancy between the mass density obtained from direct observations of gas and stars (the visible mass) and the dynamical density $\rho_\odot \approx 0.15\ M_\odot\ \mathrm{pc}^{-3}$ estimated from the model. Based on various considerations, he argued that $\approx 40\%$ of this value should be assigned to unseen (dark) matter distributed as the disk stars.

The project became the focus of renewed interest after the discovery of flat rotation curves in spiral galaxies (see Chapter 9), in terms of better and more extended data sets and also with the help of more sophisticated dynamical models. The new work led to a controversy. According to one picture,[17] the discrepancy and the conclusions noted by Oort are basically confirmed. In a similar study based on a different approach,[18] a significant discrepancy is not found, and the dynamical estimate suggests a relatively light disk with no appreciable presence of invisible (dark) matter.

The resolution of the controversy came from the analysis of fresh new data from *Hipparcos* (see also Section 2.3), which proved convincingly that there is no significant mass discrepancy in the disk in the solar neighborhood.[19] In Fig. 8.3 we illustrate the distribution of the vertical velocities for a sample of stars studied by means of the new *Hipparcos* data. The apparently negative outcome of this project that has attracted the attention of many scientists over a period of about 70 years should not be misunderstood. The negative result demonstrates that in some cases, especially those in which a measured mass

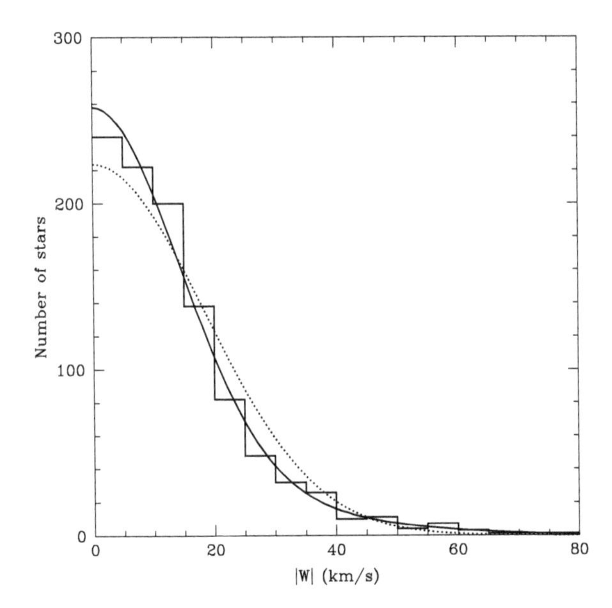

Figure 8.3 Vertical motion distribution of K giant stars in the solar neighborhood (a sphere of 100 pc radius around the Sun). The observed velocity distribution (histogram) is compared to a two-component Gaussian fit (solid line; velocity dispersions $\sigma_{z1} \approx 14$ km s^{-1} and $\sigma_{z2} \approx 28$ km s^{-1}) and a single-component Gaussian (dotted line; velocity dispersion $\sigma_z \approx 18$ km s^{-1}). The simpler one-component model proves to be insufficient for the purpose of measuring the density distribution of the stars in the dynamical model. The other components of the velocity dispersion are reported to be $\sigma_\theta = 23$ km s^{-1} and $\sigma_r = 34$ km s^{-1}. (From: Holmberg, J., Flynn, C., "The local surface density of disc matter mapped by Hipparcos," 2004. *Mon. Not. Roy. Astron. Soc.*, **352**, 440; in the article additional details are given that better identify the sample of stars used in the figure and the related fit.)

discrepancy is within a factor of two, we should be ready to scrutinize critically data and models before rushing to draw our conclusions about the presence of dark matter. Furthermore, finding a relatively light disk with no evidence for dark matter (in the disk) is a very important result by itself, because it suggests that the dark matter present (as the study of rotation curves proves) is not in the form of a thin dark disk, but instead of a roundish dark halo. In turn, the end of this controversy provides one of the strongest empirical indications that dark matter is likely to be collisionless (nondissipative), which naturally fits in with the current interpretation in terms of so-far undetected exotic particles (the picture of dark matter made of Weakly Interacting Massive Particles, WIMPs).

The value of Oort's project is further emphasized by the fact that a similar long-term investigation has been undertaken for the study of external galaxies:

the Disk Mass Project has already produced interesting results, which generally point to the existence of disks somewhat lighter (i.e., submaximal; see Chapter 9) than expected.[20]

8.3 The Isothermal Self-Gravitating Slab

In the studies referred to in the previous section, the models developed to describe the vertical equilibrium in the solar neighborhood include the presence of various classes of stars and of the interstellar medium. They also include effects related to the inhomogeneities in the plane of the disk and to the restoring terms in the vertical acceleration (for stars outside the plane of the disk) that are not directly associated with the local disk density. A detailed discussion of such complex modeling is especially required because the mass discrepancy suspected initially was not large.

In concluding this chapter, we briefly describe a simple idealized model, known as the isothermal self-gravitating slab, to illustrate only one aspect of the problem, that is, the role of self-consistency. In this sense, the model is instructive, but should not be viewed as adequate for a direct confrontation with the data.

Consider a flat, infinite distribution of stars assumed to be homogeneous in the (x, y) plane and inhomogeneous in the vertical direction. Let $\rho = \rho(z)$ be the mass density, assumed to be symmetric with respect to the equatorial plane $z = 0$, so that $\rho(z) = \rho(-z)$, that is, the density is an even function of the vertical coordinate. The density of stars is imagined to be characterized by a peak value $\rho_0 = \rho(0)$ and be a monotonic decreasing function of the distance $|z|$ from the equatorial plane. Such a geometrical configuration is often called a slab model. We assume that the slab is fully self-gravitating, that is, it generates a gravitational potential $\Phi(z)$ that is related to the density distribution by the Poisson equation

$$\frac{d^2\Phi}{dz^2} = 4\pi G\rho. \tag{8.6}$$

For the distribution of the vertical component of the stellar velocities we assume an isothermal Maxwellian (see also Subsection 1.4.1)

$$f = \frac{\rho_0}{\sqrt{2\pi}c} \exp\left(-\frac{E}{c^2}\right), \tag{8.7}$$

where $E = v_z^2/2 + \Phi(z)$. Here ρ_0 and c are constants. The quantity c determines the width of the Maxwellian and represents the vertical velocity dispersion. By definition, integrating the distribution function over the v_z velocity space gives the dependence of the density distribution on the vertical coordinate, implicitly as a function of Φ:

$$\rho(z) = \int_{-\infty}^{\infty} f dv_z = \rho_0 \exp\left[-\frac{\Phi(z)}{c^2}\right]. \tag{8.8}$$

To be consistent with the notation introduced earlier in this section, we set the condition $\Phi(0) = 0$, so that indeed ρ_0 represents the value of the density on the equatorial plane. The same relation (8.8) between density and potential would be obtained by integrating the condition of hydrostatic equilibrium $(1/\rho)dp/dz + d\Phi/dz = 0$ (see Subsection 4.4.1) for a fluid characterized by equation of state $p = c^2\rho$.

Eliminating the density from Eqs. (8.6) and (8.8) we find that self-consistency requires the solution of the second order nonlinear ordinary differential equation

$$\frac{d^2\Phi}{dz^2} = 4\pi G\rho_0 \exp\left(-\frac{\Phi}{c^2}\right), \tag{8.9}$$

under the boundary conditions $\Phi(0) = 0$ and $(d\Phi/dz)(0) = 0$ on the $z = 0$ plane.

In many cases, to solve a nonlinear differential equation we would have to resort to a numerical integration. Here we may take advantage of the analogy between the form of this equation and that of the equation of the motion determining a one-dimensional orbit $x(t)$ of a particle in a given potential $U(x)$, by considering the change of variables $(\Phi, z) \to (x, t)$. In other words, we may easily integrate once and construct an energy-like integral by multiplying (see Subsection 6.4.1) both sides of Eq. (8.9) by the pseudo-velocity $d\Phi/dz$.

The problem can be conveniently cast in dimensionless form, by introducing two scales associated with the dimensional constants ρ_0 and c. In particular, we can refer to a dimensionless binding potential $\psi \equiv -\Phi/c^2$ and a dimensionless vertical coordinate $\zeta \equiv z/z_0$, with the scale z_0 defined by

$$z_0^2 \equiv \frac{c^2}{2\pi G\rho_0}. \tag{8.10}$$

Then, in dimensionless form the self-consistency condition (8.9) becomes

$$\frac{d^2\psi}{d\zeta^2} = -2\exp(\psi), \tag{8.11}$$

to be solved under the boundary conditions $\psi(0) = 0$ and $(d\psi/d\zeta)(0) = 0$. This equation, multiplied by the pseudo-velocity $d\psi/d\zeta$, can be integrated to yield

$$\frac{1}{2}\left(\frac{d\psi}{d\zeta}\right)^2 = \mathcal{E} - 2\exp(\psi) = 2[1 - \exp(\psi)], \tag{8.12}$$

where the dimensionless energy-like constant $\mathcal{E}$ has been set equal to 2 in order to satisfy the adopted boundary conditions. The last equation is easily integrated once more, by recalling the properties of the hyperbolic cosine function. Returning to the dimensional form, we thus obtain

$$\rho(z) = \rho_0 \frac{4}{[\exp(-z/z_0) + \exp(+z/z_0)]^2} = \rho_0 \exp\left[-\frac{\Phi(z)}{c^2}\right]. \tag{8.13}$$

Three final remarks can be made on this solution. The relation between surface density Σ and volume density is

$$\Sigma = \int_{-\infty}^{\infty} \rho dz = 2\rho_0 z_0. \tag{8.14}$$

The other remarks refer to the two natural asymptotic limits

$$\Phi \sim 2\pi G\Sigma|z|, \quad z \gg z_0, \tag{8.15}$$

consistent with a derivation from the Gauss theorem, and

$$\Phi \sim \frac{1}{2}\left(\frac{d^2\Phi}{dz^2}\right)_0 z^2 = \frac{1}{2}(4\pi G\rho_0)z^2, \quad z \ll z_0. \tag{8.16}$$

If we combine Eq. (8.14) with the definition of Eq. (8.10), we obtain

$$z_0 = \frac{c^2}{\pi G\Sigma}, \tag{8.17}$$

consistent with the heuristic argument reported at the beginning of Section 8.2.

Notes

1 For example, see Wolf, M. 1906. *Astron. Nachr.*, **170**, 211 (in German).

2 In particular, in the frequently cited paper Zwicky, F. 1933. *Helv. Phys. Acta*, **6**, 110 (in German), distances are reported for eight nearby clusters of galaxies and one group of galaxies: the Virgo cluster is listed at 6×10^6 light-years and the Coma cluster at 45×10^6 light-years. The estimates reported in Tables 1 and 2 by Zwicky are based on Hubble, E., Humason, M. L. 1931. *Astrophys. J.*, **74**, 43.

3 On the one hand, an effort was made at developing comprehensive catalogues; see Abell, G. O. 1958. *Astrophys. J. Suppl.*, **3**, 211. On the other hand, great attention was given to their properties, especially in view of an adequate cosmological context; see Bahcall, N. A. 1977. *Annu. Rev. Astron. Astrophys.*, **15**, 505.

4 Zwicky, F. 1933. op. cit. focused on the dynamics of the Coma cluster. Smith, S. 1936. *Astrophys. J.*, **83**, 23 studied the Virgo cluster. Zwicky, F. 1937. *Astrophys. J.*, **86**, 217 gives a summary of methods and results.
5 Smith, S. 1936. op. cit.
6 Hubble, E. 1926. *Astrophys. J.*, **64**, 321.
7 Actually Smith made his dynamical estimate of the mass of the Virgo cluster by means of some assumptions on galaxy orbits in the outer regions of the cluster.
8 The issue of the possible presence of rotation, that is, of significant amounts of angular momentum, in clusters of galaxies is to be considered as an open question. Here we list only some of the relevant references on this topic. Gregory, S. A., Tifft, W. G. 1976. *Astrophys. J.*, **205**, 716 argued for the presence of systematic rotation in the Coma cluster. Dressler, A. 1981. *Astrophys. J.*, **243**, 26 studied the dynamics of the highly flattened cluster A2029, finding that it shows no signs of large-scale rotation. In a survey of 899 Abell clusters, Hwang, H. S., Lee, M. G. 2007. *Astrophys. J.*, **662**, 236 found tentative evidence of systematic rotation for only 12 clusters. On the side of X-ray investigations, Bianconi, M., Ettori, S., Nipoti, C. 2013. *Mon. Not. Roy. Astron. Soc.*, **434**, 1565 studied the problem of the detectability of rotation in the ICM of clusters.
9 Sarazin, C. L. 1988. *X-Ray Emissions from Clusters of Galaxies*. Cambridge University Press, Cambridge, UK. Rosati, P., Borgani, S., Norman, C. 2002. *Annu. Rev. Astron. Astrophys.*, **40**, 539.
10 Schindler, S., Hattori, M., Neumann, D. M., Böhringer, H. 1997. *Astron. Astrophys.*, **317**, 646.
11 With an additional caveat, that gravitational lensing analyses are known to be affected by the so-called mass-sheet degeneracy, because they are sensitive to density gradients rather than to absolute values of the density distributions.
12 De Boni, C., Bertin, G. 2008. *Il Nuovo Cimento B*, **123**, 31.
13 Ettori, S., De Grandi, S., Molendi S. 2002. *Astron. Astrophys.*, **391**, 841.
14 For example, see the article by Bonamigo, M., Grillo, C., et al. 2018. *Astrophys. J.*, **864**, id.98.
15 Oort, J. H. 1932. *Bull. Astron. Inst. Neth.*, **6**, 249. A summary of the work carried out on this program is given by Oort, J. H. 1965. In *Galactic Structure*, eds. A. Blaauw, M. Schmidt. University of Chicago Press, Chicago, p. 455.
16 Kapteyn, J. C. 1922. *Astrophys. J.*, **55**, 302; Jeans, J. H. 1922. *Mon. Not. Roy. Astron. Soc.*, **82**, 122.
17 Bahcall, J. N. 1984. *Astrophys. J.*, **276**, 156 and 169; Bahcall, J. N. 1984. *Astrophys. J.*, **287**, 926; Bahcall, J. N., Flynn, C., Gould, A. 1992. *Astrophys. J.*, **389**, 234.
18 Kuijken, K. H., Gilmore, G. 1989. *Mon. Not. Roy. Astron. Soc.*, **239**, 571, 605, and 651; 1991. *Astrophys. J. Lett.*, **367**, L9. Similar conclusions had been reached by other scientists; see Crézé, M., Robin, A. C., Bienaymé, O. 1989. *Astron. Astrophys.*, **211**, 1, and references therein.
19 Crézé, M., Chereul, E., Bienaymé, O., Pichon, C. 1998. *Astron. Astrophys.*, **329**, 920; Holmberg, J., Flynn, C. 2000. *Mon. Not. Roy Astron. Soc.*, **313**, 209; Holmberg, J., Flynn, C. 2004. *Mon. Not. Roy. Astron. Soc.*, **352**, 440.
20 Verheijen, M. A. W., Bershady, M. A., et al. 2004. *Astron. Nachr.*, **325**, 151; Bershady, M. A., Martinsson, T. P. K., et al. 2011. *Astrophys. J. Lett.*, **739**, id.L47. As part of the same project, the galaxy UGC 463 was studied by Westfall, K. B., Bershady, M. A., et al. 2011. *Astrophys. J.*, **742**, id.18.

9
The Discovery of Dark Halos around Spiral Galaxies

Earlier in this book we highlighted the important role of 21-cm radio observations in measuring the kinematics of spiral galaxies. In particular, at the beginning of Chapter 3 we mentioned single-dish observations (with the pioneering studies at Dwingeloo) and interferometric observations (especially the initial work of WSRT) that led to the first systematic investigations of the rotation curves of spiral galaxies (see Subsection 3.1.1). Radio observations were accompanied by optical observations, initially based on the study of emission lines produced in the interstellar medium (mostly in the regions of star formation, called HII regions). In sum, the data support the picture that in regular, normal spiral galaxies, except for deviations associated with spiral arms, bars, warps, lopsidedness, and other features, the overall character of the kinematics of the disk can be described in terms of a mean axisymmetric rotation around the center, as captured by the fluid model presented in Subsection 4.4.2. In general, the rotation is differential, that is, the angular velocity changes with radius (usually in a monotonic declining way). The disks of spiral galaxies are generally cool, in the sense that the mean rotation is dominant and random motions (stellar velocity dispersion or turbulent motions of the interstellar medium) can be considered to be small (see Section 6.4).

To appreciate the difficulty in making accurate and reliable measurements of the rotation curves much should be written and discussed. Given the introductory character of this book, we will not address the art of constructing the rotation curves from the available data. Such art is based on experience and deep understanding and is often underestimated even by dynamicists active in the field. Here we only wish to make a few comments that best apply to radio investigations using the 21-cm line of the atomic hydrogen. For an extended source such as a spiral galaxy, if it is not spatially resolved, the Doppler effect on the selected source line typically generates a double-horned observed line, which is the natural signature of the overall rotation state of

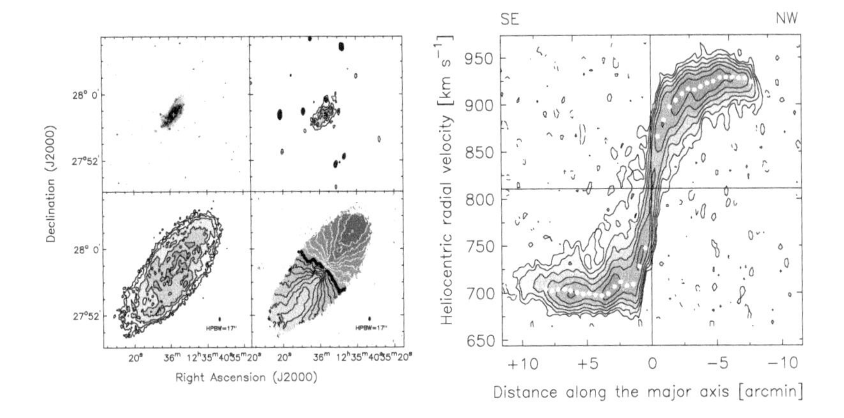

Figure 9.1 HI study of NGC 4559. Small frames to the left: optical image (top left panel), radio continuum (top right panel), column density contours (bottom left), and velocity field at 17″ resolution (bottom right panel). In the velocity field, contours are separated by 15 km s^{-1}. The scale factor is 1′ = 2.8 kpc. A small cross indicates the kinematical center of the galaxy. Big frame to the right: position–velocity diagram at 26″ resolution along the major axis (V_{sys} = 810 km s^{-1}). The white dots show the (projected) rotation curve. (From: Barbieri, C. V., Fraternali, F., et al., "Extra-planar gas in the spiral galaxy NGC 4559," 2005. *Astron. Astrophys.*, **439**, 947; reproduced with permission © ESO.)

the disk (one horn corresponds to the receding part of the galaxy, the other to the approaching part, for the velocity component along the line of sight). The structure of these global line profiles already provides significant insights. Yet, the derivation of an accurate and reliable rotation curve requires the inspection and the modeling of a complex multidimensional data set, that is, the collection of spatially resolved line shifts and line widths (derived from the so-called data-cube, requiring an evaluation of the line profile corresponding to each spatially resolved element; see also Subsection 1.3.3). In practice, the intermediate tools used by the astronomer to derive the kinematical information for the extended source comprise intensity maps (HI column density contours), isovelocity contours (i.e., maps that display the points in the sky where the measured line-of-sight velocity takes a given value), position–velocity diagrams that exhibit the intensity of the signal corresponding to a certain velocity as a function of position along a given line across the source, and the detailed channel maps (intensity distributions at a given line-of-sight velocity). For 21-cm observations of the galaxy NGC 4559, some of these quantities are illustrated in Fig. 9.1 and in Fig. 9.2 (where the term *radial velocity* stands for line-of-sight velocity).

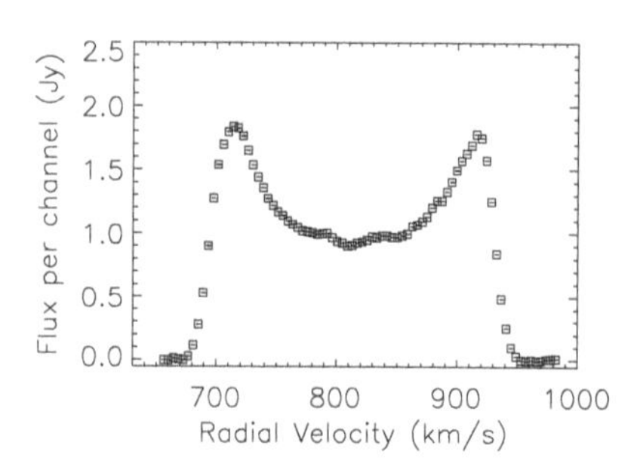

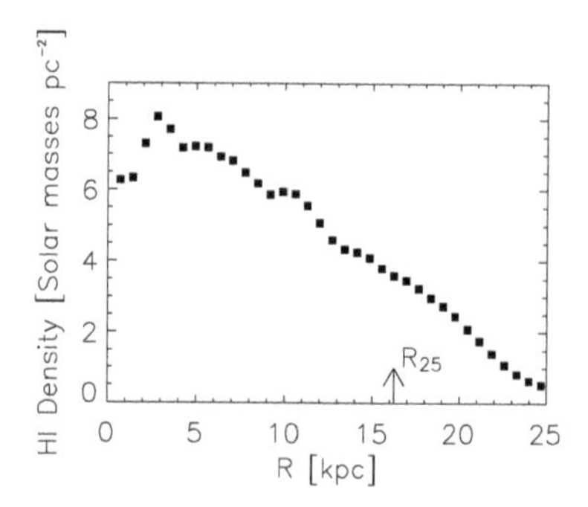

Figure 9.2 Global HI line profile (left) and column density radial profile (right) for NGC 4559 (from: Barbieri, C. V., Fraternali, F., et al., "Extra-planar gas in the spiral galaxy NGC 4559," 2005. *Astron. Astrophys.*, **439**, 947; reproduced with permission © ESO). Note the double-horned structure of the global line profile.

The variety of features that are observed have led to major discoveries about the structure of the disk.[1] Under simple and reasonable assumptions, one of the most natural and useful tools developed to bridge the gap between data and model curves is the so-called tilted-ring model of a disk.[2] A lesson that would emerge clearly from a closer inspection of the tools and methods that we have just superficially mentioned is that accuracy in the measurement of a rotation curve is generally best obtained for moderately inclined galaxies (with respect to the line of sight) and for regions of the disk not too close to the galaxy center.

In this chapter we will focus our attention on describing how the study of the rotation curves of spiral galaxies led to the surprising discovery of dark matter halos. Starting in the 1970s, especially because of the growing evidence accumulating from WSRT, clues for the existence of dark matter halos became stronger and stronger. As a result of the excitement produced by the clues that were being gathered, some confusion lingered until in the mid-1980s decisive evidence was eventually obtained (Section 9.1). Once the need for a significant contribution to the gravitational field by a dark component is recognized, we should determine the relative contribution of dark and visible matter to the total gravitational field. This step is usually called decomposition of a rotation curve (see Section 9.2), and still remains largely ambiguous. In a conservative approach, the role of dark matter is thought to be minimal and becomes dominant only in the outer parts of the galaxy. Such maximum disk solution (Subsection 9.2.1) has often been argued to be preferable, but there are indications that it is not strictly followed by real galaxies. After many decades of investigations on this topic there remain several unexplained empirical aspects about the measured gravitational field ("conspiracy") and rather large uncertainties about the relative amounts of dark and visible matter ("degeneracy"), which we will examine in Subsection 9.2.2. Touching upon a nontrivial dynamical issue, we will continue the chapter by raising the question of self-consistency in relation to the decomposition of rotation curves (Section 9.3). Finally, in Section 9.4 we will summarize two dynamical arguments that go beyond the direct inspection of the properties that characterize the observed basic state of spiral galaxies and call for the presence of a dark halo as a solution to otherwise unexplained stability properties of galaxy disks.

9.1 Flat Rotation Curves

That some nebulae are rotating was recognized well before the advent of radio astronomy and even before the discovery of galaxies.[3] Of course, the study of rotation curves and, especially, their interpretation came much later. As noted in Chapter 3, the first measurements of the rotation curve of M31 with the Dwingeloo radio telescope[4] gave indications that at large distances from the center the rotation velocity remains higher than expected (see Fig. 3.1). However, such behavior was not taken to be a definite sign of a major problem. In fact, the standard reference for the mass model of our Galaxy was the Schmidt model,[5] according to which the rotation curve of the Milky Way Galaxy starts to decline in a Keplerian way ($V \sim r^{-1/2}$) just beyond the orbit of the Sun. The Schmidt model was still in use in many articles of the mid-1970s, even though at that time evidence that external galaxies are characterized by flat, nondeclining rotation curves was getting stronger and stronger.

In general, the rotation curve $V(r)$, as measured by radio and optical spectroscopic observations, starts with an approximately linear increase in the central regions and then, at larger radii, tends to flatten out, within a variety of detailed shapes.[6] Surprisingly, evidence for a Keplerian decline in the outer parts of rotation curves has long been sought but has not been found. Instead, it was soon noted that some galaxies exhibit a flat rotation curve even on enormous scales; in particular, the optical rotation curves of NGC 801 and UGC 2885 remain approximately flat at ≈ 200 km s^{-1} from 4 to 60 kpc and at ≈ 300 km s^{-1} from 2 to 80 kpc, respectively.[7]

As anticipated in Chapter 4, Eq. (4.11), in the case of cool disks (for which the pressure term associated with the velocity dispersion of the stars or the turbulent motions of the gas can be ignored) the measurement of the rotation curve provides a direct measurement of the gravitational acceleration in the plane of the disk:

$$\frac{V^2}{r} = \frac{d\Phi}{dr}. \tag{9.1}$$

The central linear rise ($V \sim r$ at small radii), which implies an approximately solid-body rotation, is not surprising for a gravitational acceleration $d\Phi/dr$ generated by a regular distributed mass (e.g., see the discussion given in Section 7.5 and at the end of Section 8.1). What caught the attention of astronomers was the absence of a Keplerian decline in the outer parts and the existence of a wide radial range in which rotation curves are approximately flat. As to the implied existence, amount, and distribution of dark matter (often referred to as missing mass or undetected matter) the arguments were various and sometimes not as clear-cut as desired.

9.1.1 Initial Discussions

In this respect, one interesting discussion was given by Morton Roberts.[8] Roberts made several points. (1) Optically measured rotation curves frequently refer to only the inner parts of the bright stellar disk, reaching out to $\approx R_{opt}/3$ or $\approx R_{opt}/2$ [in terms of the optical radius of the disk; for an exponential disk with exponential length h (see Section 6.3), a reasonable definition of the optical radius is $R_{opt} = 4.5h$]. (2) Radio rotation curves often extend well beyond the bright optical disk, that is, beyond R_{opt} (see Fig. 9.3), but may suffer from sensitivity problems. (3) In many cases the radio rotation curves remain flat in most of the outer regions, out to the last measured point. (4) The behavior often observed in the outer parts suggests that significant amounts of mass are located at large radii (although the total mass of the galaxy does not increase to the last measured point by a large factor). (5) Thus, in many galaxies it appears, as noted earlier,[9] that the mass distribution does not reflect the light distribution (which for the disk is generally declining exponentially; see Section 6.3 and Fig. 9.3). (6) The shapes of the rotation curves of different galaxies appear to be different; in particular, some well-studied galaxies, such as M81 and M83, appear to have a declining rotation curve at large radii. The interesting points (4) and (5) were basically made in several papers in the 1970s and early 1980s.[10] Others also focused on point (6), suggesting a correlation between the shape of the rotation curve and galaxy category along the Hubble sequence or galaxy luminosity.[11] Point (1) turned out to be a serious problem and the acquisition of accurate and radially extended radio rotation curves eventually led to the decisive evidence for the existence of dark matter halos.

Some confusion in the interpretation of the observations should be traced to the improper replacement of Eq. (9.1) by

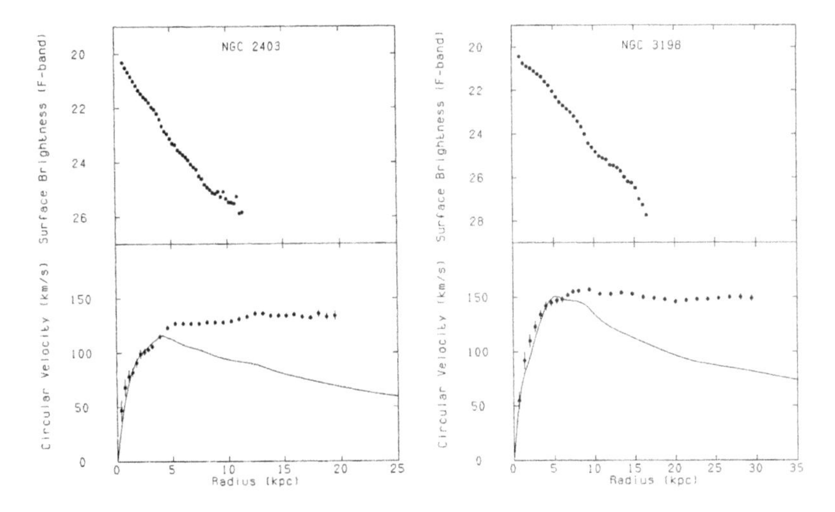

Figure 9.3 The exponential luminosity profile of the spiral galaxies NGC 2403 (left) and NGC 3198 (right), and the related radio-determined rotation curves (from: Sancisi, R., and van Albada, T. S., "HI rotation curves of galaxies," 1987. In IAU Symposium: *Dark Matter in the Universe*, eds. J. Kormendy and G. R. Knapp. Reidel, Dordrecht, The Netherlands, p. 67; © International Astronomical Union). Solid curves in the lower panels show the rotation curves expected for models with a constant (maximum) mass-to-light ratio for the stellar disk and no dark matter.

$$\frac{V^2}{r} = \frac{GM(r)}{r^2}, \tag{9.2}$$

where $M(r)$ represents the amount of matter enclosed in a sphere of radius r. Such expression for the gravitational acceleration is correct and justified only if the density distribution generating the gravitational field is characterized by spherical symmetry. Obviously, if we take literally this last equation instead of the correct Eq. (9.1), we conclude that (i) in the outer parts, where, in the absence of dark matter, $M(r) \sim M_{tot}$, with M_{tot} the total mass of the galaxy, we would expect a Keplerian decline $V \sim r^{-1/2}$, and, in turn, (ii) a flat rotation curve requires $M(r) \propto r$. Because the luminosity of the (star-dominated) disk decreases exponentially, the observation of a flat rotation curve would thus require the mass not to be distributed in proportion to the disk luminosity profile, and dark matter must be present. In reality, the relation may be justified only asymptotically $V^2 \sim GM(r)/r$, at large radii.[12] Additional undesired ambiguities affect a proper interpretation of the rotation curve and reduce the impact on conclusions about the existence of dark matter; in particular, if the galaxy under consideration is complex (i.e., it has a prominent bulge or a dynamically important gas component or features such as strong deviations from

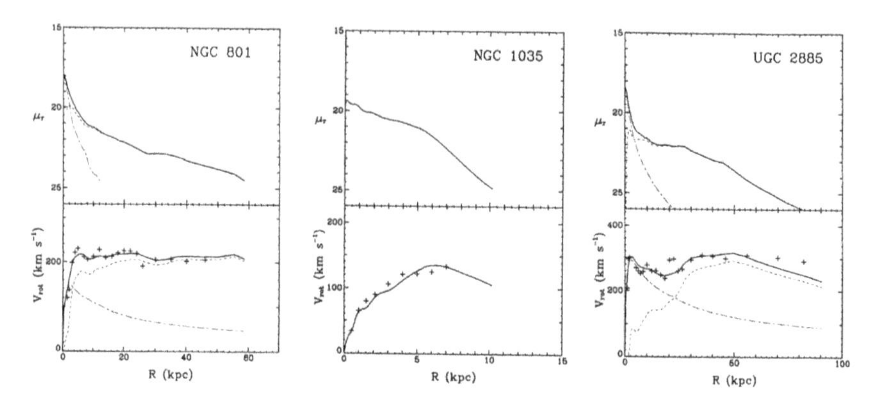

Figure 9.4 Three cases (one small and two large Sc spirals) from a study of 37 late-type galaxies (from: Kent, S., "Dark matter in spiral galaxies: I. Galaxies with optical rotation curves," 1986. *Astron. J.*, **91**, 1301; reproduced by permission of the AAS) for which a fit to the available optical rotation curve is made with a bulge-disk decomposition without a dark halo. In each galaxy, the disk and the bulge mass-to-light ratios are treated as constant free parameters (NGC 1035 is taken as having no bulge).

axisymmetry) the discussion includes factors that can be combined to produce effects that make the interpretation in terms of dark matter confused.

In the early 1980s, a provocative test on four galaxies, NGC 4378 (Sa), NGC 7217 (Sb), NGC 598 (M33, Sc), and NGC 7793 (Sd),[13] proved that the arguments in favor of the existence of dark matter had to be formulated in a more convincing and decisive way. The flat character of the rotation curve available at that time is especially marked in NGC 4378 and NGC 7217; for NGC 4378 the data-points exhibit an approximately flat rotation curve from 2 to 23 kpc. The test showed that, by converting for each galaxy the observed brightness distribution (photometric data-points) into a density distribution under the assumption of a constant mass-to-light ratio M/L and by computing the associated gravitational field $d\Phi/dr$ of Eq. (9.1) by means of the Poisson equation (following the path outlined in Subsection 9.2.1), an expected rotation curve $V_{vis} = \sqrt{rd\Phi/dr}$ associated with the visible mass is able to match the observed optical rotation curve $V(r)$ without resorting to any dark matter contribution, for reasonable values of the M/L parameter. For the early-type galaxy NGC 4378 the analysis required the decomposition of the photometric data-points and the corresponding visible mass into bulge and disk contributions (and thus the related decomposition of the rotation curve; see Section 9.2). The result was confirmed by a similar test performed on 37 galaxies with optical rotation curves (see Fig. 9.4).[14]

9.1.2 Decisive Evidence of the Presence of Dark Halos

Eventually it was realized that a clear-cut interpretation of the rotation curve is best obtained by considering galaxies with a pure disk (many Sc spiral galaxies are essentially bulgeless), not too small (small galaxies tend to have low levels of rotation and often contain large amounts of gas), with intermediate inclination with respect to the line of sight (so as to avoid difficulties associated with the study of disks that are close to being viewed face-on or edge-on), and without major asymmetries (bars, prominent grand-design spiral arms, warps, lopsidedness, or other features). For the purpose of obtaining decisive evidence for the existence of dark matter, radio observations are those that must be used because in many cases they can take advantage of the presence of a regular HI disk extending out well beyond the bright optical stellar disk. Finally, in the modeling process, Eq. (9.1) should be used and the generally inapplicable Eq. (9.2) should be avoided. The decisive evidence was indeed obtained by careful and detailed studies of galaxies such as the Sc spiral galaxies NGC 2403 and NGC 3198, illustrated in Fig. 9.3 (for NGC 2403, see also Fig. 6.3).[15] For NGC 3198 the visible mass is distributed as a thin disk and the surface brightness profile is very well fitted by an exponential (see Section 6.3); the measured radio rotation curve extends out to 11 exponential scale lengths ($\approx 2R_{opt}$), with error bars ≈ 5 km s^{-1}, and remains flat at ≈ 150 km s^{-1} from 2 to 11 exponential scale lengths.[16]

9.1.3 Pure Stellar Disk and Spherical Dark Matter Halo

Consider a spiral galaxy with no bulge in which the mass of the interstellar medium is negligible, that is, a pure stellar disk. In the thin disk approximation we convert the observed disk surface brightness profile $I(r)$ (photometric data-points) into a disk mass density profile $\Sigma(r) = (M/L)I(r)$ by means of a constant mass-to-light ratio M/L. From the physical point of view we are considering the case of homogeneous stellar populations, for which the observed luminosity is directly proportional to the underlying stellar mass distribution (see Section 1.3). It may be convenient to take a model in which the disk is infinitesimally thin, so that the Poisson equation relating the mass distribution to the mean potential Φ is

$$\nabla^2\Phi = 4\pi G\Sigma(r)\delta(z) = 4\pi G\left(\frac{M}{L}\right)I(r)\delta(z), \tag{9.3}$$

where $\delta(z)$ is the standard Dirac δ-function and, as usual, z denotes the vertical coordinate, with the disk placed at $z = 0$. This equation can be solved for Φ.[17] In other words, for an axisymmetric disk in which the density distribution Σ

depends only on the radial distance r there are closed expressions that allow us to compute the mean potential associated with a given density distribution and the gravitational acceleration on the disk plane $d\Phi/dr$. Then we can also compute the expected rotation curve $V_d = \sqrt{rd\Phi/dr}$ associated with the visible disk. Therefore, for a given $I(r)$ the shape of the curve is determined uniquely, but its scale remains undetermined (by a factor $\sqrt{M/L}$). For the special case in which $I(r)$ is strictly exponential, there is an explicit solution for V_d in terms of standard modified Bessel functions[18]; after an initial monotonic rise at small radii, the function V_d reaches a peak at ≈ 2.2 exponential scale lengths and then declines monotonically (turning to the asymptotic behavior $V_d \sim r^{-1/2}$ characteristic of Keplerian circular orbits). At 11 exponential scale lengths the value of V_d is $\approx 1/2$ the peak value attained at ≈ 2.2 exponential scale lengths. This last remark makes it clear that in the case of NGC 3198 mentioned earlier the only way to explain the observed rotation curve, flat out to 11 exponential scale lengths, is to assume that there is an additional contribution to the gravitational field due to undetected matter.

As described in Section 8.2, the failure of Oort's and Bahcall's projects in detecting a mass discrepancy in the solar neighborhood proved that, at least for our Galaxy, there is no evidence for dark matter distributed in a thin disk. Therefore, it is natural to imagine that dark matter, if present (as proved by the study of radially extended rotation curves), should be distributed in a round halo. Physically, this is consistent with the modern view that dark matter is likely to consist of collisionless particles (WIMPs; see the beginning of Chapter 5). In the simplest case, if we consider a strictly spherical dark matter halo characterized by cumulative mass $M_{DM}(r)$, its contribution to the gravitational field would be precisely

$$\frac{d\Phi_{DM}}{dr} = \frac{GM_{DM}(r)}{r^2}. \tag{9.4}$$

Therefore, asymptotically, at large radii, a flat rotation curve V would require $M_{DM}(r) \sim r$, that is, a dark matter density distribution $\rho_{DM}(r) \sim r^{-2}$. As we briefly mentioned in Subsection 8.1.3, this density profile is the signature of the self-gravitating isothermal sphere. For simplicity, we may thus refer to the approximate analytic expression that we used in the context of the dynamics of clusters of galaxies and adopt the following form for the dark matter density distribution:

$$\rho_{DM} = \rho_{DM}^{(0)} \frac{1}{[1 + (r/r_{DM})^2]}, \tag{9.5}$$

where r_{DM} is the core radius of the dark matter distribution.

9.2 Decomposition of an Observed Rotation Curve

Having discovered that spiral galaxies are embedded in dark halos, we may proceed to the task of determining, case by case, how much dark matter is required and how it should be distributed to explain the observed rotation curve. This process is called decomposition of an observed rotation curve. In general, we wish to identify the different contributions of the visible mass (stellar disk, gaseous disk, and bulge) and of the dark matter halo. Much like in the way we did in Subsection 4.4.1, for the interpretation of the hydrostatic equilibrium condition of the ICM in clusters of galaxies, and described in Subsection 8.1.3 to assess the relative concentrations of dark and visible mass in the clusters, we can proceed as follows.

We rewrite Eq. (9.1) with the right-hand side listing explicitly the contributions of the different components:

$$\frac{V^2}{r} = \frac{d\Phi_{d\star}}{dr} + \frac{d\Phi_{gas}}{dr} + \frac{d\Phi_b}{dr} + \frac{d\Phi_{DM}}{dr}. \tag{9.6}$$

The sum of the first three terms on the right-hand side represents the gravitational acceleration on the galactic plane produced by the visible mass $d\Phi_{vis}/dr$, whereas the last term identifies the contribution to the field associated with dark matter. In practice, astronomers simplify the problem of connecting the data with the dynamical model, by limiting the decomposition to the introduction of only few parameters. In particular, for the stellar disk the contribution $d\Phi_{d\star}/dr$ is calculated by inserting in the Poisson equation the star density $\Sigma_\star$ obtained from the photometric profile of the disk by means of a constant mass-to-light ratio $(M/L)_{d\star}$. For the gas the contribution $d\Phi_{gas}/dr$ is calculated by inserting in the Poisson equation the gas density Σ_{gas} obtained from the measured HI atomic hydrogen density profile, multiplied by a constant factor f (constrained only in part by the observations) to include the presence of molecules and cosmological helium. Therefore, the disk contribution to the gravitational field is $d\Phi_d/dr = d\Phi_{d\star}/dr + d\Phi_{gas}/dr$. Finally, the bulge contribution $d\Phi_b/dr$ is calculated by inserting in the Poisson equation the bulge density distribution obtained from the photometric profile of the bulge by means of a constant mass-to-light ratio $(M/L)_b$ (which may be different from $(M/L)_{d\star}$, because bulge and stellar disk are likely to have a different formation history). As a result, the gravitational field associated with the visible mass can be reconstructed from the data, but it depends on the specification of three parameters [$(M/L)_{d\star}$, f, and $(M/L)_b$] that are only in part constrained by the observations. For the dark matter contribution, the gravitational field is often modeled in terms of the field associated with the density distribution given in Eq. (9.5), which has a simple analytical expression and depends on two parameters (the central density $\rho_{DM}^{(0)}$

and the core radius r_{DM} of the dark matter distribution); alternatively, a dark matter density profile is often assumed following the suggestions obtained from cosmological simulations of structure formation (see Subsection 10.3.1).[19]

Traditionally, that is, following the general notation adopted in the last five decades, the decomposition is expressed by referring to

$$V^2 = V_{d\star}^2 + V_{gas}^2 + V_b^2 + V_{DM}^2, \tag{9.7}$$

obtained from Eq. (9.6) by multiplication by r, with the obvious notation $V_i^2 \equiv r d\Phi_i/dr$. In addition, the decomposition is generally illustrated by plotting the data-points of the observed rotation curve $V(r)$, together with the model curves $V_i(r)$; in these plots, it should be recalled that the various contributions $V_i(r)$ to the model rotation curve $V_{mod}(r)$ are meant to be added in quadrature, following Eq. (9.7).

One point that is interesting to note and further demonstrates the importance of not using Eq. (9.2) to estimate the gravitational field associated with a disk density distribution is the study of the gas contribution V_{gas}^2. In many cases the density distribution of the gas is nonmonotonic and may even have a central hole (e.g., in galaxies such as M81). A correct calculation of the gas contribution to the gravitational field, by means of the Poisson equation applied to calculate the right-hand side of Eq. (9.1), shows that in certain cases in the central regions V_{gas}^2 can be negative, that is, it can contribute a repulsive force with respect to the center of the galaxy.[20]

In conclusion, a given set of data-points for a rotation curve $V(r)$ is studied in terms of the dynamical model Eq. (9.1), and a best-fit analysis is performed to measure the parameters that are at the basis of the fit, that is, $(M/L)_{d\star}$, f, and $(M/L)_b$ for the visible matter together with $\rho_{DM}^{(0)}$ and r_{DM} for the dark matter. If strong arguments can be put forward in favor of a specific choice of $(M/L)_{d\star}$, f, and $(M/L)_b$, the best-fit analysis has fewer free parameters, and we are basically left with the direct determination of the amount and distribution of dark matter.

9.2.1 The Maximum Disk Hypothesis

As was immediately realized after the provocative test on the four galaxies NGC 4378, NGC 7217, NGC 598, and NGC 7793 mentioned earlier in this chapter, it appears that the central parts of a rotation curve can be explained without resorting to the presence of dark matter. This preliminary conclusion turned out to be confirmed not only when the test was extended to all the available optical rotation curves but also later, in the following years, when the attention was turned to Low Surface Brightness galaxies (Subsection 6.2.2).

If, for simplicity, we refer to the decomposition of a pure stellar disk, for which the contributions of V_{gas} and V_b are negligible, we find that the rotation curve data can be fitted with a maximum value of the mass-to-light ratio $(M/L)_{d\star}$ [higher values of $(M/L)_{d\star}$ would make the contribution of the visible disk overshoot the rotation curve data-points], whereas the contribution from the dark halo emerges only at large radii. In the fit process this solution is a conservative solution, in the sense that it minimizes the amount of dark matter present. It is called the maximum disk solution and it is often argued to hold as the preferred solution, under the name of the maximum disk hypothesis.[21]

However, because the presence of a dark halo is required anyway (by the observations of the rotation curve in the outer regions), we might consider the possibility that such a halo contributes significantly also to the field in the inner regions. For a pure stellar disk, we may thus consider a value of $(M/L)_{d\star}$ smaller than that determined by the maximum disk hypothesis and argue that there is a sufficient amount of dark matter in the halo to compensate, in the inner regions, for the difference between the observed rotation curve and the expectation based on such a lighter (nonmaximum) disk.

In view of the main argument in support of the maximum disk hypothesis, that is, the attempt at minimizing the role of the dark halo, the concept can be easily generalized to the case of more complex disks. In general, the maximum disk hypothesis postulates that the rotation curve of spiral galaxies should be decomposed so as to maximize the role of the visible mass (V_{vis}), with a fit based on the three parameters $(M/L)_{d\star}$, f, and $(M/L)_b$ that makes the role of dark matter negligible inside the bright optical disk. It is commonly recognized that such maximum disk decomposition is viable also for LSB galaxies, even though for LSB galaxies the associated value of $(M/L)_{d\star}$ turns out to be generally larger than that suggested by the study of stellar populations.[22]

9.2.2 Conspiracy and Degeneracy

A key unresolved problem that still haunts us after more than a century of progress and speculations about the issue of dark matter is the apparently simple question what is dark matter made of? On various occasions, we have noted that there is a general consensus on the possibility that dark matter is made of some so-far undetected particles, in particular WIMPs. Many projects are under way with the goal of confirming this conjecture. The nature of dark matter is obviously the main open question that we would like to answer. Yet, there are other very important unresolved questions that observations raise and these are captured by two catchwords, *conspiracy* and *degeneracy*, that are often used in this research area.

If two different components (the baryonic matter that makes the visible mass and the dark matter that we have been forced to invoke to explain the observed rotation) contribute to the gravitational field in galaxies, why should they cooperate precisely so as to determine, in many cases, such flat rotation curves? We recall that, for NGC 3198, the measured radio rotation curve extends out to 11 exponential scale lengths ($\approx 2R_{opt}$), with error bars ≈ 5 km s^{-1}, and remains flat at ≈ 150 km s^{-1} from 2 to 11 exponential scale lengths. If we take the point of view of the maximum disk hypothesis, the unexpected cooperation becomes even more evident. In fact, in the inner regions the value of ≈ 150 km s^{-1} would be largely determined by the visible disk, whereas in the outer regions it is basically determined by the dark halo alone. Why should there be such fine tuning? In the absence of an answer, this problem has been dubbed a conspiracy. There are other aspects of the problem of dark matter that also fall under the same issue of fine-tuning, that is, conspiracy. For example, in Subsection 1.2.3 we mentioned the important Tully–Fisher relation, which states that the total optical luminosity L of a spiral galaxy correlates with the scale of its rotation V as $L = aV^4$. If the luminosity is determined only by visible matter and the velocity scale is determined by both visible and dark matter, why should a relation of this type hold, with such small dispersion? The stellar component and the dark halo must have had a very different formation and evolution history; why should they conspire to produce together such a tight empirical relation?

The other catchword, *degeneracy*, is also well exemplified by a detailed study of the disk–halo decomposition of the rotation curve of NGC 3198. If we perform a parametric decomposition for this galaxy, in which for simplicity we neglect (as is reasonably well justified by the data) the V_{gas} and V_b contributions, and adopt for the dark matter halo the simple density distribution expressed by Eq. (9.5), we can show that the rotation curve can be very well fitted by an entire strip of models in the relevant two-dimensional parameter space.[23] A standard statistical analysis shows that models identified by parameters along the strip produce fits that are all excellent and basically indistinguishable in terms of quality. At one end of the strip there is the maximum disk solution; for the models at the other end, the disk is negligible and the gravitational field is dominated by the dark halo at all radii. This is called degeneracy. How can we tell which solution actually corresponds to the structure of NGC 3198 and other galaxies?

9.3 Self-Consistent Decomposition

In Section 8.3 we discussed the structure of the self-gravitating isothermal slab. Earlier, in Section 8.1 we mentioned the isothermal sphere, for which the density distribution is approximately described by Eq. (9.5). In these cases, the hypothesis of isothermality corresponds to assuming that the relevant system under investigation is associated with a Maxwellian distribution function; for the slab, see Eq. (8.7). The argument of the Maxwellian is the specific energy $E = v^2/2 + \Phi$, where Φ represents the gravitational potential describing the mean field within the galaxy.

Here we wish to point out that a self-consistent decomposition of a rotation curve based on the hypothesis that the galaxy disk is embedded in an isothermal dark matter halo cannot be performed by simply superposing the contribution of the disk $d\Phi_d/dr$ on the gravitational field associated with the spherical density distribution of Eq. (9.5). The reason is that the potential Φ that appears in the argument of the Maxwellian is the total gravitational potential. For a pure stellar disk embedded in an isothermal dark halo, we can write $\Phi = \Phi_{d\star} + \Phi_{DM}$ following the notation used earlier in this chapter. The halo density distribution is associated with the halo distribution function by the relation $\rho_{DM} = \int f_{DM} d^3v$ and self-consistency requires the potential associated with the dark matter distribution to obey the Poisson equation

$$\nabla^2 \Phi_{DM} = 4\pi G \rho_{DM} = 4\pi G \int f_{DM} d^3v. \tag{9.8}$$

But now, on the right-hand side the source of the gravitational field cannot be strictly spherically symmetric, because f_{DM} depends on the total gravitational potential (Φ is partly determined by the potential of the disk $\Phi_{d\star}$, which is only axisymmetric and certainly is not spherically symmetric). In simpler words, the fact that the isothermal halo hosts an axisymmetric disk with given properties implies that, by self-consistency, the isothermal halo must be flattened by the presence of the disk, especially at small radii, and can keep its desired spherical symmetry only asymptotically at large radii. The solution for Φ_{DM} (and consequently for ρ_{DM}) of the self-consistency requirement set by Eq. (9.8) in the presence of an exponential disk is technically difficult, but has been worked out in detail;[24] here we just mention that the resulting self-consistent decomposition of the rotation curve has been found to alleviate the problem of degeneracy (see Fig. 9.5), in the sense that it has led to the identification of best-fit models associated with disks that are slightly lighter than those expected from the maximum disk hypothesis.[25]

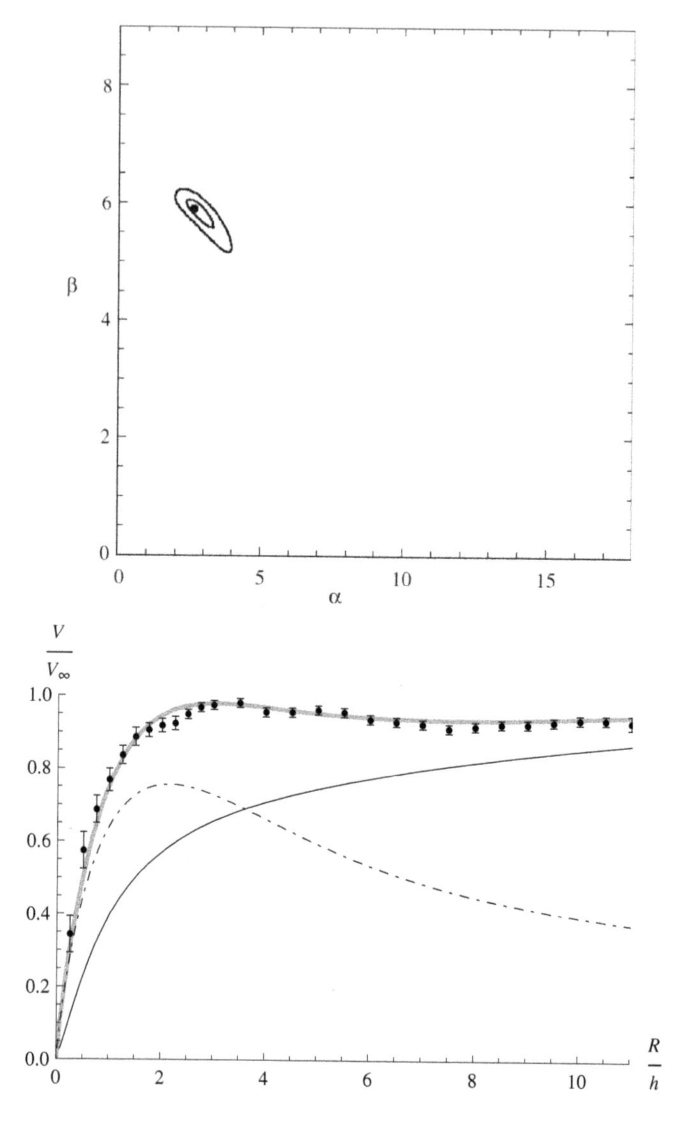

Figure 9.5 Removal of the disk–halo degeneracy by a self-consistent decomposition of NGC 3198. (Top) Contours of 68% and 95% confidence regions in the (α, β) parameter plane. Here the parameter α can be seen as the dimensionless measure of the central dark matter density scale and β is the dimensionless weight of the stellar disk as measured by the mass-to-light ratio $(M/L)_{d\star}$; the maximum-disk solution would be characterized by $\beta \approx 10$. (Bottom) Best-fit disk–halo decomposition. (From: Amorisco, N. C. and Bertin, G., "Self-consistent nonspherical isothermal halos embedding zero-thickness disks," 2010. *Astron. Astrophys.*, **519**, id.A47; reproduced with permission © ESO).

9.4 Dynamical Arguments

If we change the characteristics of the basic state of a dynamical system we also change the way it responds to perturbations and, in general, its stability properties. This obvious statement could be exemplified by referring to several completely different contexts, from the notes produced by a musical instrument, to the behavior of geophysical flows or to the study of the stability of plasma equilibrium configurations.

Therefore, if, as we are forced to do as a consequence of the discovery of dark halos, we change the picture of the general structure of spiral galaxies, we must be ready to change our views about the way galaxy disks would respond to given perturbations. In other words, dark halos not only change the structure of disk galaxies, but also change their dynamics.

As we have often noted throughout this book (see Section 6.3, Section 8.3, and Subsection 4.4.2), the symmetries of the basic state of galaxy disks, with or without the presence of an embedding dark halo, are axisymmetry, with respect to an axis passing through the center and orthogonal to the disk, and reflection symmetry, with respect to the equatorial plane of the disk. This suggests that we may discuss separately, for small (linear) perturbations, the properties of waves characterized by different azimuthal wavenumber m in a standard Fourier analysis [with elementary dependence $\exp(im\theta)$]. Perturbations with $m \neq 0$ break the axisymmetry of the basic state. Similarly, for small (linear) perturbations we can study separately the properties and the stability of even (density) or odd (bending) waves, where the mathematical terms *even* and *odd* refer to the reflection symmetry with respect to the horizontal equatorial plane of the disk. Note that in this framework, density waves are related to the spiral morphology of galaxy disks; $m = 2$ modes would be associated with bars or two-armed spiral structure. Bending waves would break the reflection symmetry of the disk; in particular, $m = 1$ bending waves would correspond to a warp of the disk characterized by a change of sign of the deviation from the plane of the disk on opposite sides of the galaxy disk.

Here we recall two specific investigations that show how deep the connection between the discovery of dark halos and our understanding of the dynamics of galaxy disks is.

9.4.1 Bars

A very interesting conjecture was put forward at the very beginning of the study of the rotation curves of spiral galaxies.[26] Based on a collection of some numerical simulations and of previously published investigations, including the study of the ellipsoidal figures of equilibrium that we mentioned in Subsection 3.5.1, it was argued that axisymmetric self-gravitating rotating systems that are dynamically too cold should be unstable and change their symmetry because of the development of unstable bar modes. If we refer to the quantity appearing in Eq. (6.21), the proposed criterion of instability would state that axisymmetric systems characterized (approximately) by $K_{int}/|W| < 0.36$ should be unstable against bar modes.[27] To some extent, this general conjecture was confirmed later by a detailed global modal analysis aimed at understanding the morphology of spiral galaxies.[28]

In relation to the content of the present chapter, the very interesting part of the argument is that the authors of the conjecture noted that our Galaxy, which is basically axisymmetric, in spite of some clear deviations from axisymmetry, would be a counterexample to the proposed criterion, because its disk is too cold. The authors then argued that one way to reconcile the symmetry of our Galaxy with the stability criterion just proposed would be to imagine that our Galaxy is embedded in a round, slowly rotating dark halo, with mass within the scale of the bright optical disk similar to the mass of the disk. In the global energy budget, such a halo would not increase the amount of kinetic energy under the form of ordered motions K_{ord} but would increase the magnitude of the gravitational energy $|W|$, thus lowering the value of $K_{ord}/|W| = 1/2 - K_{int}/|W|$; a sufficiently massive slowly rotating or nonrotating dark halo could thus increase the value of $K_{int}/|W|$ so as to make it consistent with the conjectured stability criterion, against the apparent violation that would be judged by considering the coldness of the disk (in the absence of the dark halo). In conclusion, on pure theoretical grounds and, curiously, without a rigorous proof, this dynamical argument anticipated the modern, current picture of the structure of dark halos.

9.4.2 Warps

Since the 1950s it has been realized that the HI disk of our Galaxy is warped, with a dominant $m = 1$ component.[29] The phenomenon clearly called for an investigation of the stability of a thin disk against bending waves. It was soon realized that self-gravitating disks should be stable with respect to bending perturbations.[30] However, the initial studies had been performed before the discovery of dark halos in spiral galaxies and thus no halo was present in the dynamical system for which the conclusion on the stability of bending waves had been reached. To understand the existence of the warp in our Galaxy it was then proposed that our disk would suffer from tidal interactions, but a quantitative analysis of this possibility seemed to indicate that the available tidal interactions (in particular, the interaction with the Large Magellanic Cloud) are too weak to justify the amplitude of the observed distortion.

The problem of galaxy warps gained renewed attention after a radio investigation of five edge-on galaxies showed that warps are the rule rather than an exception.[31] The galaxies that were examined[32] are NGC 5907, NGC 4565, NGC 4244, NGC 4631, and NGC 891. The most pronounced and unambiguous warp was found in NGC 5907. Curiously, these observations took place right at the time when the phenomenon of flat rotation curves was becoming a key issue in extragalactic astronomy, leading the way to the discovery of dark halos.

One possibility that emerged as a way to interpret the frequently observed warps is that they result naturally from disk–halo interaction, so that galaxy disks would flap much like a flag in the wind, given the relative motion between the slowly rotating or nonrotating halo and the fast-rotating disk. This would be some sort of an analogue, in the galactic context, of instabilities (in particular, Kelvin–Helmholtz and two-stream instabilities) that are well-known in fluid dynamics and plasma physics.[33]

Notes

1 For example, see Sancisi, R., Fraternali, F., Oosterloo, T., van der Hulst, T. 2008. *Astron. Astrophys. Rev.*, **15**, 189.
2 In which the radio-emitting disk is imagined as being made of concentric rings, characterized by given rotation velocity and freely oriented with respect to the observer; see Rogstad, D. H., Lockhart, I. A., Wright, M. C. H. 1974. *Astrophys. J.*, **193**, 309. A modern advanced tool for the extraction of rotation curves from spectroscopic observations, based on the general idea of the tilted-ring model, has been produced by Di Teodoro, E. M., Fraternali, F. 2015. *Mon. Not. Roy. Astron. Soc.*, **451**, 3021.
3 For the Sombrero galaxy NGC 4594; see Slipher, V. M. 1914. *Bull. Lowell Obs.*, **62, II**, n. 12.
4 van de Hulst, H. C., Raimond, E., van Woerden, H. 1957. *Bull. Astron. Inst. Neth.*, **14**, 1.
5 Schmidt, M. 1965. In *Galactic Structure*, eds. A. Blaauw, M. Schmidt. University of Chicago Press, Chicago, p. 513.
6 For example, see Fig. 4 in Casertano, S., van Gorkom, J. H. 1991. *Astron. J.*, **101**, 1231.
7 Rubin, V. C., Burstein, D., Ford, W. K., Thonnard, N. 1985. *Astrophys. J.*, **289**, 81. The optical rotation curves of NGC 801 and UGC 2885 are reported later in this chapter, in Fig. 9.4.
8 Roberts, M. S. 1976. *Comm. Astrophys.*, **6**, 105.
9 Freeman, K. C. 1970. *Astrophys. J.*, **160**, 811; especially in relation to the radio rotation curve of NGC 300. This is the same article that was quoted in Chapter 6 in relation to the photometry of galaxy disks. See also the discussion of the kinematics of NGC 2403 in Shostak, G. S. 1973. *Astron. Astrophys.*, **24**, 411.

10 In particular, see Faber, S. M., Gallagher, J. S. 1979. *Annu. Rev. Astron. Astrophys.*, **17**, 135; Bosma, A. 1981. *Astron. J.*, **86**, 1791 and 1825.
11 Rubin, V. C., Ford, W. K., Thonnard, N., Burstein, D. 1982. *Astrophys. J.*, **261**, 439. See also Rubin, V. C. 1987. In *Dark Matter in the Universe*, eds. J. Kormendy, G. R. Knapp. Reidel, Dordrecht, The Netherlands, p. 51.
12 If light traces mass, Eq. (9.2) is justified for locations well outside the region where most of the visible matter is observed. If, in addition to the visible matter, we consider the presence of dark matter, it is justified only if the dark halo is spherically symmetric.
13 Kalnajs, A. 1983. In *Internal Kinematics and Dynamics of Galaxies*, ed. E. Athanassoula. Reidel, Dordrecht, The Netherlands, p. 87. In the test, Kalnajs made use of optical rotation curves.
14 Kent, S. M. 1986. *Astron. J.*, **91**, 1301.
15 van Albada, T. S., Sancisi, R. 1986. *Phil. Trans. Roy. Soc. London A*, **320**, 447.
16 van Albada, T. S., Bahcall, J. N., Begeman, K., Sancisi, R. 1985. *Astrophys. J.*, **295**, 305.
17 For example, see Eqs. (14.26), (14.29), and (14.30) in Bertin, G. 2014. *Dynamics of Galaxies*, 2nd ed. Cambridge University Press, New York.
18 For example, see Eq. (14.32), in Bertin, G. 2014. op. cit.
19 Navarro, J. F., Frenk, C. S., White, S. D. M. 1996. *Astrophys. J.*, **462**, 563; Chemin, L., de Blok, W. J. G., Mamon, G. A. 2011. *Astron. J.*, **142**, 109.
20 For the galaxy NGC 4559, illustrated in Figs. 9.1 and 9.2 of this chapter, this feature is clearly exhibited by the dashed line in the rotation curve decomposition shown in Fig. A.1 of the article Barbieri, C. V., Fraternali, F., et al. 2005. *Astron. Astrophys.*, **439**, 947.
21 van Albada, T. S., Sancisi, R. 1986. op. cit.
22 Sancisi, R. 2004. In *Dark Matter in Galaxies*, eds. S. D. Ryder, D. J. Pisano, M. A. Walker, K. C. Freeman. Publ. Astron. Soc. Pacific, San Francisco, p. 233. In the abstract of this paper, Sancisi also states: “For any feature in the luminosity profile there is a corresponding feature in the rotation curve and vice versa.”
23 See Fig. 15 in Amorisco, N. C., Bertin, G. 2010. *Astron. Astrophys.*, **519**, id.A47; see also Figs. 17 and 18 in the same article.
24 By iteration, by seeding the right-hand side of Eq. (9.8) with the potential Φ^0_{DM} of the fully self-gravitating regular isothermal sphere.
25 Amorisco, N. C., Bertin, G. 2010. op. cit.
26 Ostriker, J. P., Peebles, P. J. E. 1973. *Astrophys. J.*, **186**, 467.
27 Actually the criterion formulated by Ostriker and Peebles was based on the complementary parameter $t = K_{ord}/|W|$. Under equilibrium conditions, the virial theorem requires $t = 1/2 - K_{int}/|W|$.
28 Bertin, G., Lin, C. C. 1996. *Spiral Structure in Galaxies: A Density Wave Theory*. MIT Press, Cambridge, MA, and references therein.
29 It is not the purpose of this book to address this issue in detail and thus most of the vast literature on this topic will be ignored. The interested reader may consult Chapter 19 of the monograph Bertin, G. 2014. *Dynamics of Galaxies*, 2nd ed. Cambridge University Press, New York, NY, and the many references given there.
30 Hunter, C., Toomre, A. 1969. *Astrophys. J.*, **155**, 747.
31 Sancisi, R. 1976. *Astron. Astrophys.*, **53**,159. In the summary of the article Sancisi notes that “three of these systems are fairly isolated in the sky.”
32 See the galaxies illustrated on p. 25 in Sandage, A. 1961. *The Hubble Atlas of Galaxies*. Publ. 618, Carnegie Institution of Washington, Washington, DC.
33 Bertin, G., Mark, J. W.-K. 1980. *Astron. Astrophys.*, **88**, 289.

10
The Cosmological Context

This final chapter will give us the opportunity to comment on some modern developments in the study of the problem of dark matter in astrophysics and to address several important points that relate the problem of dark matter (as developed so far in Part II) to the cosmological context. As we noted for several other topics on previous occasions, we do not intend to cover thoroughly the issues that we will raise, but rather we aim to make the reader aware of the existence of very interesting questions, some of which would deserve an entire monograph on their own.

The discovery of dark matter in galaxies marks a turning point in astrophysics especially because of its conceptual relevance to cosmology. The first sparks of evidence for the existence of dark halos starting with the early 1970s literally ignited a fire of enthusiastic reactions from cosmologists and generated feedback among the various scientific communities involved, much like the recent detection of gravitational waves has produced a general enthusiastic surge of interest in a variety of topics (particularly, black holes and compact stars) in physics and astrophysics.

In Section 10.1 we will describe some of the main findings in the study of dark halos in elliptical galaxies, for which progress has been made with some delay with respect to the study of spiral galaxies, simply because ellipticals lack the direct dynamical estimator associated with rotation curves. In Section 10.2 we will briefly comment on the relevance of galactic dark halos to the determination of the cosmological parameters (the cosmological parameters were introduced in Section 2.2 and then mentioned in Section 3.2 in relation to the discovery of the Cosmic Microwave Background radiation); the main conclusion that we wish to emphasize here is that, although conceptually the discovery of dark matter halos changes our view of the universe, in practice the dark matter contained in galaxies does not change the value of the key cosmological parameters significantly. In Section 10.3 we will refer to some modern simulations of structure formation. This is a vast and fast-evolving field of research that ultimately focuses on the very interesting issues that define the formation and the evolution of galaxies. Note that, in turn, the focus of this short primer is largely on the structure and dynamics of galaxies. As a result, we will not consider in detail and examine critically the line of research of cosmological simulations. Still, some description and some comments are in order, especially in relation to the structure of dark halos formed as a result of cosmological evolution. Then, in Section 10.4 we will recall the potential of *Gaia* observations in clarifying the structure and dynamics of the nearby universe. In Section 10.5 we will raise a question that is currently attracting the interest of dynamicists, for which nontrivial modeling and observations are required, that is, the presence of dark matter in globular clusters. In Section 10.6 we will touch on the role of gravitational lensing as diagnostics of dark matter for galaxies and clusters of galaxies in the distant universe; in fact, it would be impossible to conclude this short introduction to astrophysical dynamics and dark matter without giving adequate credit to gravitational lensing as a key tool in modern astrophysics. Finally, in Section 10.7 we will outline the main steps of an attempt (MOND, MOdified Newtonian Dynamics), started in the early 1980s and still continuing with some success, which basically follows the cautionary remarks by Richard Feynman quoted at the beginning of Chapter 6 ("Of course we cannot prove that the law here is precisely inverse square, only that there is still an attraction ..."): given the fact that so far we have been unable to identify the constituents of dark matter, the attempt explores the possibility that the law of gravitation requires a modification on the scales of galaxies and beyond and that dark matter does not exist.

10.1 Dark Matter Halos in Elliptical Galaxies

The main message of Chapter 9 is that, after a period of more than a dozen years of interesting investigations and discussions on the properties of rotation curves, in the mid-1980s we finally obtained convincing evidence of the presence of dark halos in spiral galaxies. What about elliptical galaxies? At this point it would be natural to expect that, in general, ellipticals should also be embedded in dark halos. This expectation is especially natural for those who believe that, as a rule, ellipticals are born as a result of mergers of spiral galaxies.[1] But how could we measure the amount and distribution of dark matter in elliptical galaxies? The following incomplete and simplified description that we give as an answer to this question should make it clear that a deep discussion would require stellar-dynamical modeling tools beyond the goals of the present book.[2]

Unfortunately, elliptical galaxies generally lack a straightforward tracer of their gravitational field, as the rotation curve offers for spiral galaxies [see Eq. (9.1)]. Normally, they do not host regular thin disks of atomic hydrogen as spiral galaxies do.[3] As stellar systems, their kinematical state is revealed by the study of absorption lines associated with the stellar atmospheres, which is a difficult task and generally leads to kinematical data-points with large error bars. Surprisingly, in the mid-1970s, when the first kinematical measurements became available, it was realized that bright ellipticals do not rotate even when they are flat; the kinetic energy of the stars is in the form of random motions, rather than ordered mean motions. Therefore, typical optical spectroscopic studies of bright ellipticals measure velocity dispersion (along the line of sight) profiles and such kinematical measurements extend only out to approximately the effective radius (i.e., not as far out as desired, for an adequate sampling of the gravitational field, with respect to the bright optical galaxy; for the definition of the effective radius, see Subsection 1.2.3).

10.1.1 Some Properties of Collisionless Stellar Systems

The absence of rotation in flat galaxies reminded astronomers that ellipticals are collisionless stellar systems. We briefly addressed the important concept of collisionality in Section 1.4 in relation to the study of globular clusters, in Section 5.3 in relation to the solar wind, and in Sections 6.2 and 6.3 in the context of spiral galaxies. At variance with collisionally relaxed systems, collisionless stellar systems generally have an anisotropic velocity dispersion. In other words, at a given location the distribution of the star velocity vectors may be characterized by different dispersions when different directions are considered. An empirical statement that exemplifies this concept for the solar neighborhood is given in the caption to Fig. 8.3.

Given the goals of this introductory book, we did not introduce properly the concept of distribution function for the description of collisionless stellar systems and their dynamics, although we touched on some basic concepts such as relaxation and frequently referred to the Maxwell–Boltzmann distribution function. Before continuing with our short discussion of the presence of dark matter in elliptical galaxies, we may just state without derivation some properties of collisionless equilibria as described by means of a one-star distribution function $f = f(\vec{x}, \vec{v})$, the probability of finding a star at a given position $\vec{x}$ with given velocity $\vec{v}$ (phase space).

The general equation that governs the evolution of f is

$$\frac{Df}{Dt} = \frac{\partial f}{\partial t} + \vec{v} \cdot \frac{\partial f}{\partial \vec{x}} - \frac{\partial \Phi}{\partial \vec{x}} \cdot \frac{\partial f}{\partial \vec{v}} = 0, \tag{10.1}$$

where standard notation is used for gradients in position and velocity space. This equation is sometimes called the collisionless Boltzmann equation or the Vlasov equation.

The distribution function can be normalized to the mass density ρ in the sense that the relation between density and distribution function is $\rho = \int f d^3 v$. The field $\vec{u}$ of mean motions (flow velocity) associated with f is defined as the average in velocity space of $\vec{v}$

$$\vec{u} = \frac{1}{\rho} \int \vec{v} f d^3 v. \tag{10.2}$$

Then, the velocity dispersion tensor is defined as

$$\sigma_{ij}^2 = \frac{1}{\rho} \int (v_i - u_i)(v_j - u_j) f d^3 v. \tag{10.3}$$

By taking moments of Eq. (10.1) in velocity space we obtain a hierarchy of fluid equations. Among these, we obtain equations that correspond exactly to the Euler equation for fluids Eq. (4.2), with the difference that the pressure term on the right-hand side now involves the divergence of the pressure tensor

$$p_{ij} = \rho \sigma_{ij}^2. \tag{10.4}$$

In stellar dynamics, these fluid-like equations are sometimes called the Jeans equations.

For a basic state with assigned symmetry, the general equilibrium condition to be obeyed by f is that it must be a positive-definite integral of the motion for the mean potential $\Phi(\vec{x})$ that defines the relevant stellar orbits; sometimes this statement is called the Jeans theorem.

For a spherically symmetric collisionless stellar system, a natural equilibrium distribution function is a positive definite function of the specific energy E and of the specific angular momentum squared $J^2 = r^2(v_\theta^2 + v_\phi^2)$, where we have used standard polar spherical coordinates. For such $f(E, J^2)$ no mean motions are present (i.e., $\vec{u}(\vec{x}) = 0$) and the velocity dispersion tensor is generally anisotropic, with $\sigma_{rr}^2 \neq \sigma_{\theta\theta}^2 = \sigma_{\phi\phi}^2$. As a local measure of the velocity dispersion anisotropy, we can introduce the function

$$\alpha = \alpha(r) = 2 - \frac{\sigma_{\theta\theta}^2 + \sigma_{\phi\phi}^2}{\sigma_{rr}^2}. \tag{10.5}$$

Note that the isotropic case is characterized by $\alpha = 0$. In turn, we say that the anisotropy is radially biased if $\alpha > 0$ and, conversely, tangentially biased if $\alpha < 0$.

For a nonrotating spherically symmetric collisionless stellar system, the hydrostatic equilibrium condition corresponding to Eq. (4.4) can be written as

$$\frac{1}{\rho}\frac{d(\rho\sigma_{rr}^2)}{dr} + \alpha\frac{\sigma_{rr}^2}{r} = -\frac{d\Phi}{dr}, \tag{10.6}$$

with $\alpha = \alpha(r)$ defined by Eq. (10.5).

10.1.2 Dark Matter in Ellipticals

It was immediately realized that, whereas the luminosity profile of elliptical galaxies appears to have a universal character (embodied by the $R^{1/4}$ law suggested by Gérard de Vaucouleurs), some ellipticals have steeper and others have flatter velocity dispersion profiles. Then it was soon noted that, with respect to the rather rapidly declining velocity dispersion profile observed in NGC 3379,[4] a flatter profile may be generated by either the presence of a dark halo or the presence of tangentially biased velocity dispersion anisotropy. The role of anisotropy in changing the velocity dispersion profile was indeed studied by means of Eq. (10.6).[5]

A physically justified picture of galaxy formation, by violent collisionless collapse, has led the way to interpret why for bright elliptical galaxies the luminosity profile is universal. The key process involved is that of incomplete violent relaxation,[6] which has been taken as the basis for the construction of successful dynamical models. It leads to equilibria characterized by isotropic velocity dispersion inside the effective radius and radially biased anisotropy in the outer regions. These models have been used to decompose the observed kinematical data in terms of contributions of visible (stars) and dark matter, in analogy with the decomposition of rotation curves in spiral galaxies. Bright ellipticals with relatively large effective radii have shown evidence[7] of the presence of a dark halo with properties not far from those found in spirals. For the galaxy NGC 4472 the decomposition is illustrated in Fig. 10.1. In the left panel, the decomposition is shown in the same spirit of the decomposition expressed by Eq. (9.7): the letters L and D mark the curves $V_L = \sqrt{r(d\Phi_L/dr)}$ and $V_D = \sqrt{r(d\Phi_D/dr)}$ called circular velocities of the luminous and dark components, respectively. It should be emphasized that in the model there is no rotation: the circular velocities are given only as a way to represent the strength of the gravitational field associated with the two components. Two important points should be noted in the figure. The first point is that the superposition of the contributions of luminous and dark matter conspires to produce a total

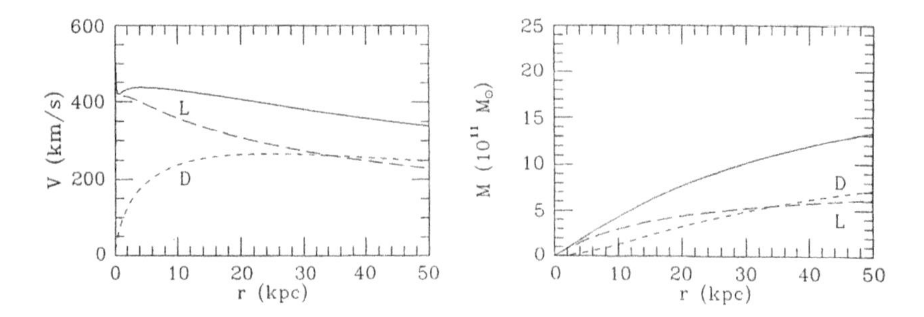

Figure 10.1 Intrinsic properties for the best-fit two-component model of NGC 4472, based on models that incorporate the picture of formation under incomplete violent relaxation (from: Saglia, R. P., Bertin, G., and Stiavelli, M., "Elliptical galaxies with dark matter: II. Optimal luminous-dark matter decomposition for a sample of bright objects," 1992, *Astrophys. J.*, **384**, 433; reproduced by permission of the AAS): for the dark (D) and the luminous (L) component, circular velocity profile decomposition (left) and cumulative mass decomposition (right).

gravitational field associated with a rather flat, only slightly declining, circular velocity curve, at ≈ 400 km s^{-1}, out to 50 kpc ($\approx 4R_e$, where the best-fit model to the available kinematical data has been extrapolated). The second important point, illustrated by the right panel, is that inside 50 kpc the mass associated with dark matter is of the same order of magnitude as that associated with the visible matter.

For small ellipticals the situation is less clear. Some ellipticals of intermediate size and luminosity (in particular, NGC 3379) do not show evidence of significant amounts of dark matter.

Given the difficulties affecting stellar-dynamical measurements and some controversies affecting their modeling, with the advent of X-ray astronomy in the 1970s, great hopes were initially placed in obtaining information relevant to the problem of dark matter from X-ray investigations of those ellipticals that are embedded in a hot X-ray-emitting plasma, typically characterized by temperatures below 1 keV. In this case, X-ray observations are used as diagnostics of the gravitational field much like they are used for clusters of galaxies (i.e., by application of the condition of hydrostatic equilibrium; see Subsection 4.4.1). Unfortunately, acquiring accurate measurements of temperature and temperature gradients in the hot interstellar medium with excellent spatial resolution turned out to be much more difficult in the galactic context than anticipated. Such difficulties in the observations have often led to nonconvincing results.

Even now, in modern investigations, X-ray diagnostics is better suited for the study of clusters of galaxies than for the study of individual galaxies.[8]

In stellar systems, planetary nebulae stand out because they are very bright, are associated with clear and sharp emission lines, and, as natural products of stellar evolution, give a fair representation of the hosting stellar population. At the end of the century, it was realized that they could thus be used as excellent kinematical tracers well outside the effective radius to infer the amount and distribution of dark matter in early-type galaxies. Great progress has been made in this line of research, also with the use of dedicated telescopes. In particular, it was confirmed that NGC 3379 is likely to lack significant amounts of dark matter.[9] Similar studies have made use of globular clusters as discrete kinematical tracers, because they are found in large numbers in early-type galaxies and are relatively easy to identify. One difficulty with the use of globular clusters for the purpose of measuring the gravitational field of the galaxy where they are hosted is that they have complex dynamics, related partly to their formation and partly to their interactions with the stars of the galaxy in which they are orbiting.

Strong gravitational lensing has emerged as an important diagnostic tool for distant ellipticals. Some results will be mentioned in Section 10.6.

10.2 Impact of Galaxy Halos on the Cosmological Density Parameter

In Section 2.2 we introduced the key cosmological parameters H_0, Ω_m, and Ω_Λ. In particular, the density parameter measures the current mean matter-density of the universe ρ_0 (including dark matter), in terms of the critical density, following the definition $\Omega_m = \rho_0/\rho_{crit} = 8\pi G\rho_0/(3H_0^2)$; the mean density ρ_0 refers to an average over a sufficiently large scale, of the order of 100 Mpc. Before the discovery that the universe is currently accelerating, one cosmological model often considered for its internal elegance and simplicity was the Einstein–de Sitter model, characterized by $\Omega_\Lambda = 0$, $\Omega_m = 1$, and flat geometry.[10] Unfortunately, all the data collected initially, on the distribution of visible matter, to estimate ρ_0 pointed to values well below the critical density, by orders of magnitude, so the conclusion was that in practice $\Omega_m \ll 1$.

With the discovery of dark matter halos in galaxies, cosmological models required full revision. The acquired decisive evidence that in addition to the visible matter dark matter exists reopened the case of the estimate of Ω_m. However, it was soon realized that the impact of galactic dark matter halos on the value of Ω_m is not significant. The argument is very simple. The structure of dark matter halos around spiral galaxies required by the existence of flat rotation curves suggests that the mass inside a spherical volume of radius r should increase as $M(r) \sim r$. In principle, this would allow for a large increase of mass to be associated with an individual galaxy; in practice, because the best kinematical tracers at our disposal (in particular, the atomic hydrogen) extend out at most to a few optical radii, the mass that observations indicate we should assign to a given galaxy can increase at most by only a few times the value of the visible mass, thus leaving unchanged the order of magnitude of Ω_m based on galaxies. A related question often raised and addressed by several projects is whether we can detect the end (i.e., the outer boundary) of dark matter halos in galaxies. This problem still remains largely unresolved.[11]

Following the discovery of dark matter halos in galaxies, the study of the dynamics of clusters of galaxies at the end of the last century and at the beginning of the present century confirmed the idea that large amounts of mass are stored in clusters of galaxies. These investigations were carried out by means of improved X-ray observations (see Section 8.1) and weak-lensing analyses (see Section 3.4 and Section 10.6). They confirmed old indications[12] that $\approx 85\%$ of the total mass of clusters of galaxies is made of dark matter, with a significant contribution to Ω_m, so that the value of the matter density parameter should fall in the range $\Omega_m \approx 0.2$–0.3. In the meantime, the discovery of the acceleration of the universe pointed to $\Omega_\Lambda \approx 0.7$, so the current concordance model is still characterized by flat geometry, with $\Omega_m + \Omega_\Lambda = 1$.

In modern cosmology the dark matter first discovered in galactic halos thus plays a fundamental role. Cosmologists tend to make dark matter the primary actor of cosmic evolution, to the extent that many cosmological simulations of structure formation describe the evolution of a universe made of only dark matter, to which visible matter is added in a second stage to illustrate the properties of the universe directly observed with our telescopes. Still we should not forget that what we know about cosmic evolution comes from the empirical information provided by the visible matter. In Section 6.2 we briefly referred to some problems at the frontier of the dynamics of galaxies that comprise the frequency of occurrence of substructures, the different characteristics of small galaxies and small stellar systems, and the discovery of disks of satellites around the Milky Way Galaxy and nearby galaxies. In Subsection 9.2.2, we noted some fine-tuning problems associated with the presence of dark matter, in particular for the well-known Tully–Fisher relation. The interplay between visible and dark matter in well-studied stellar systems in the nearby universe raises key questions that cosmology will eventually have to face and to answer.

10.3 Numerical Simulations of Structure Formation in the Universe

Cosmological studies have been converging in the direction of a standard model (the ΛCDM model, that is, a cold dark matter model that now recognizes the presence of cosmic acceleration as discovered at the end of last century, associated with the Λ parameter; see Section 2.2). Given the successes obtained in the study of the CMB radiation (see Section 3.2) and the availability of major surveys on the distribution and the physical properties of galaxies in the universe (such as the 2-degree Field Galaxy Redshift Survey and the Sloan Digital Sky Survey), cosmologists aim to trace the properties of the observed universe back to the very mild fluctuations present at the beginning of the Big Bang. The impossibility in following the nonlinear evolution of the growth of structures and substructures analytically from small initial perturbations and the great progress in computers have thus encouraged the use of major *ab initio* numerical simulations to deal with the problem of structure formation from assigned initial conditions.

In this context, one major study is the *Millennium Simulation.*[13] It studies the collisionless evolution of the dark matter component, which is the dominant mass component, by using $N \approx 10^{10}$ simulation particles from redshift $z = 127$ to the present in a (periodic) cubic region $500h^{-1}$ Mpc on a side, where $1 + z$ is the expansion factor of the universe relative to the present and h is Hubble's constant in units of 100 km s^{-1} Mpc^{-1}. The code used was a special version of GADGET2,[14] based on an improved method to evaluate gravitational interactions within the category of tree codes.[15] Individual simulation particles have mass of $\approx 10^9\ M_\odot$. In practice the simulation considered $h = 0.73$, $\Omega_m = 0.25$, and $\Omega_\Lambda = 0.75$. The relation with the final galaxy distribution is obtained by applying a semi-analytic model, that is, a simplified simulation of the galaxy formation process, where the star formation and its regulation by feedback processes are parameterized in terms of simple analytic physical models; in practice, simple physical prescriptions are applied to the results of the collisionless simulation to estimate the distribution and the properties of galaxies.

To bridge the gap between the structure formation computed from the evolution of the collisionless dark matter and the observed distribution of galaxies, simulations should include a direct study of the evolution of the visible matter, which is a most challenging task. The ambitious *Illustris Project*[16] addresses such a goal, with the capability of reproducing observed features such as the cosmic star formation rate density, the galaxy luminosity function, the observed mix of early-type and late-type galaxies, the general properties of rotation curves of spiral galaxies, and scaling laws such as the Tully–Fisher relation. The project is based on sophisticated hydrodynamical simulations of the evolution of baryonic matter, thus incorporating directly many key physical processes that are involved in galaxy formation. In the article that introduces the project, the authors state that the simulations achieve a dark matter mass resolution of $6.26 \times 10^6 M_\odot$ and an initial baryonic mass resolution of $1.26 \times 10^6 M_\odot$; for comparison, some simulations have been run with dark matter only and others ("nonradiative") with reduced physics. At $z = 0$ their simulation volume contains approximately 40,000 well-resolved galaxies. New physical ingredients were later incorporated, leading to the *IllustrisTNG Project*,[17] in which the main novel feature is the capability of including magnetic fields by means of magnetohydrodynamic simulations.

All of this is very interesting and provides much food for thought. Yet, we should emphasize that insisting on developing more and more realistic simulations generally satisfies the need for having a view of evolutionary processes rather than clarifying which basic physical mechanisms are responsible for a given astrophysical phenomenon. Astrophysicists wish not only to be able to "reproduce" some observed phenomena in great detail, but also to identify the dominant elementary mechanisms that operate behind specific questions raised by the observations and to make quantitative predictions based on simple, synthetic models.

10.3.1 The Density Profile of Dark Matter Halos

One point that has raised the general interest of the astrophysical community was made on the basis of pioneering cosmological simulations performed at the end of last century, before the discovery of the cosmological acceleration. This was the finding that, on a wide range of scales, the structures formed in the universe simulated as pure collisionless dark matter appear to be characterized by a universal, homologous density distribution.[18] Once the structures are, for simplicity, described as spherically symmetric, they are associated with a density profile that can be written as

$$\rho_{NFW} = \frac{\rho_s}{(r/r_s)(1 + r/r_s)^2}. \tag{10.7}$$

The profile is characterized by two dimensional parameters, a density scale ρ_s and a length scale r_s, much like the expression that approximates the distribution of a regular isothermal sphere [see Eq. (8.5)].

The associated gravitational potential is

$$\Phi_{NFW} = -4\pi G \rho_s r_s^2 \frac{\ln(1 + r/r_s)}{(r/r_s)}. \tag{10.8}$$

With this choice the potential is finite at the origin, $\Phi_{NFW}(r = 0) = -4\pi G \rho_s r_s^2$, and vanishes at infinity. As to the associated gravitational field, from Eq. (10.7) or from Eq. (10.8), we obtain for the square of the relevant circular velocity

$$V_{NFW}^2 = \frac{GM_{NFW}(r)}{r} = r\frac{d\Phi_{NFW}}{dr} = 4\pi G\rho_s r_s^2 \left[\frac{\ln(1 + r/r_s)}{(r/r_s)} - \frac{1}{(1 + r/r_s)}\right]. \tag{10.9}$$

The circular velocity reaches its maximum, $V_{NFW}^2 \approx 0.216 \times 4\pi G\rho_s r_s^2$, at $r \approx 2.15r_s$. Curiously, the total mass, obtained by taking the limit $r \to \infty$ of the cumulative mass $M_{NFW}(r)$, diverges, whereas the gravitational energy, calculated from Eq. (6.27), is finite.

The authors of the paper that suggested Eq. (10.7) as a universal profile of dark matter halos then decided to relate their findings to the cosmological context by expressing the density scale in units of the z-dependent critical density

$$\rho_s = \delta_c \hat{\rho}_{crit}; \tag{10.10}$$

here the z-dependent critical density $\hat{\rho}_{crit} = 3H^2/(8\pi G)$ is defined in terms of the value of the Hubble constant at the redshift where the halo is located, that is, by using for the Hubble constant the expression given by Eq. (2.1). Then the authors express the length scale r_s in units of the radius r_{200} as

$$r_s = \frac{r_{200}}{c}. \tag{10.11}$$

As to the notation used, we recall that in this study the radius r_{200} (conventionally called "virial radius") is defined in such a way that the mean density associated with it is related to the z-dependent critical density in the following way:

$$200\hat{\rho}_{crit} = 3\frac{\int_0^{r_{200}} \rho(r)r^2 dr}{r_{200}^3} = \frac{M_{200}}{(4\pi/3)r_{200}^3}. \tag{10.12}$$

The two dimensionless parameters δ_c and c are called characteristic overdensity and concentration, respectively. They are mutually dependent because the definition of r_{200} requires

$$\delta_c = \frac{200}{3}\frac{c^3}{[\ln(1 + c) - c/(1 + c)]}. \tag{10.13}$$

Apparently, the 19 clumps generated as structures by the cosmological collisionless simulation mentioned at the beginning of this section, which span four orders of magnitude in mass (from the mass scale of galaxies to the mass scale of clusters of galaxies), are all well fitted by the simple profile of Eq. (10.7). The best-fit value of the concentration parameter ranges from $c \approx 20$ for the lowest mass clumps, with M_{200} applicable to galaxies, to $c \approx 5$ for structures with M_{200} applicable to clusters of galaxies. Correspondingly, the values of r_{200}

range from ≈ 200 kpc for the low mass clumps up to ≈ 4 Mpc for the highest masses.

We should recall here that significant efforts, on the theoretical and on the observational side, have been devoted to the so-called cusp problem, that is, the question of which description (not necessarily the $1/r$ behavior associated with ρ_{NFW}) would best represent the density distribution of dark matter in the vicinity of the center of dark halos.

It turned out that, as a global description of the density profiles of dark matter halos, the function[19]

$$\rho_E = \rho_S \exp\left[-\left(\frac{r}{r_S}\right)^{1/n}\right] \tag{10.14}$$

has become rather popular and is often preferred over the use of ρ_{NFW} of Eq. (10.7). This function is also characterized by two scales, ρ_S and r_S, but it requires the specification of an additional dimensionless parameter, the index n. This is the three-dimensional analogue of a function often used to describe the surface brightness profile of elliptical galaxies as a generalization of the $R^{1/4}$ law proposed by de Vaucouleurs. Of course, finding that different values of n are required for different systems would argue against the universality of the density profile of dark matter halos.

Whatever description is found to be best suited, it would be very interesting to confirm whether indeed dark matter halos are characterized by a universal density profile and, if this is the case, to find which physical mechanism is responsible for such universal behavior.[20]

10.4 *Gaia* Studies of the Near Universe

In Section 2.3 we anticipated the great potential associated with the progress in astrometric measurements in relation to dynamics, dark matter, and cosmology. As an interesting example, in Section 8.2 we recalled the decisive role of the astrometric measurements by *Hipparcos* in leading to the resolution of the controversy on the possible existence of a thin dark matter disk in our Galaxy. *Gaia* is now providing excellent measurements of the position, parallax, and annual proper motion for a very large number of stars with unprecedented astrometric precision. This will have a major impact on studies of the near universe, particularly of the structure and dynamics of our Galaxy and its neighborhood, and thus on some key issues of cosmological interest. As a currently hot research topic, we already mentioned, in Section 5.4, the important role of *Gaia* in the study of hypervelocity stars.

The connection between cosmology and the dynamics of the near universe, in particular of the Local Group of galaxies in which our Galaxy and M31 are the primary actors, is very natural. Probably, the first and most important reason to be mentioned for such connection is the fundamental task of a correct setting of the distance scales. Without elaborating further on this theme, we may just mention another issue, the classical study by Kahn and Woltjer[21] that predicted the presence of significant amounts of intergalactic matter in our Group on the basis of a very limited set of kinematical data. Even before *Gaia*, we reached the capability of measuring the velocity vectors of galaxies such as M31, thus leading to enormous progress[22] with respect to the pioneering studies of Kahn and Woltjer.

To focus on one point of great interest in this general context, here we go back to the question raised in Section 6.2, when we mentioned the possible existence of disks of satellites around nearby massive galaxies, such as M31 and NGC 5128, and in particular around our Galaxy.[23] For the Milky Way's disk of classical satellite galaxies, a recent investigation has reconsidered the overall picture in light of *Gaia*'s second data release.[24] The correlation of the orbital poles of the 11 classical satellite galaxies of our Galaxy has been confirmed. The orbital planes of 8 satellites[25] have been found to align within 20 degrees of a common direction. To measure the orbital poles, the authors of this investigation convert the heliocentric satellite positions, velocities, and proper motions to a Cartesian Galactic coordinate system centered on the Galactic center, calculate the specific angular momentum vector as the cross product of the position and velocity vectors of the satellites, and thus determine its direction. Combining the best-available proper motions substantially increases the discrepancy with the expectations from the standard ΛCDM cosmological models, given the fact that less than 1% of simulated satellite systems in *IllustrisTNG* contain seven orbital poles as closely aligned as observed.

10.5 Dark Matter in Globular Clusters?

An intriguing property that distinguishes globular clusters from dwarf galaxies is well summarized by the radius–luminosity diagram illustrated in Fig. 10.2. Here the radius R_h is the half-light radius introduced earlier in this book (e.g., see Section 6.2; this radius is sometimes denoted by R_e) as a natural linear scale for galaxies and stellar systems, and the luminosity is the absolute luminosity in the visible.[26] If we consider the range of length scales from one or a few parsecs (a natural scale for globular clusters) to 10^3 pc (so as to cover the length scale of dwarf spheroidals) and the luminosity range for M_V from 0 mag to -15 mag,

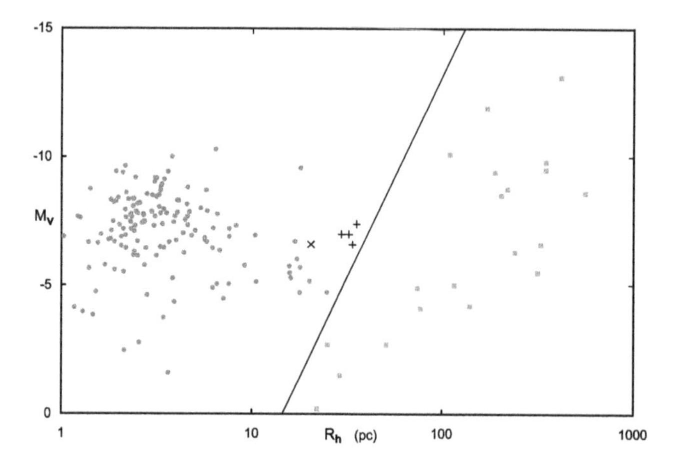

Figure 10.2 Separation between the distribution of Galactic globular clusters (filled circles) and dwarf spheroidal companions to the Galaxy (filled squares). Four extended globular clusters in M31 (plus signs) and one extended cluster in M33 (cross) are also included. The data are most incomplete in the lower right-hand corner of the diagram, that is, for the faintest and largest objects (from: van den Bergh, S., "The luminosity-diameter relations for globular clusters and dwarf spheroidal galaxies," 2008. *Mon. Not. Roy. Astron. Soc.*, **390**, L51).

we see that Galactic globular clusters and dwarf spheroidals have a not too different range in luminosities (although the luminosity distribution of dwarf spheroidals has larger dispersion), but dwarf spheroidals are distinctly more extended in radius than globular clusters are. Curiously, it is currently believed that globular clusters are basically devoid of dark matter, whereas faint dwarf spheroidals are often thought to be the objects with the largest ratio of dark to visible mass.[27] The picture of small stellar systems and satellite galaxies, in relation to their dark matter content, is further complicated by the recent studies of UDG galaxies (see Subsection 6.2.2).

It is likely that many of the unresolved issues related to the measurements of the amount and distribution of dark matter in small stellar systems result from the complexity of these systems and the lack of sound dynamical modeling tools to investigate their structure. In turn, it may well be that for globular clusters we will soon have some definite answers. To be sure, as we noted in Section 3.3, the issue of determining the possible presence of dark matter in globular clusters is made difficult by the fact that important amounts of dark remnants (white dwarfs, neutron stars, black holes) are expected to be present, especially in the central regions of the clusters, as natural products of stellar evolution. In practice, observationally, it is not easy to assess how large the mass present in the form of dark remnants is. Still, great progress has been made in understanding the internal dynamics of globular clusters, especially because of their smooth and regular observed structure and the recent observations (in particular, by *Gaia*) that have allowed us to probe the six-dimensional phase space of their stars.

10.6 Lensing Studies of Dark Matter in the Distant Universe

In Section 3.4 we described the discovery of gravitational lensing and anticipated that the phenomenon plays an important role in modern astrophysics as diagnostics of dark matter. In this context, we recall that the main idea is to model the observed characteristics of a given gravitational lensing configuration, and thus to infer back properties of the density distribution associated with the lens (the mass concentration between observer and very distant sources along a given line of sight, responsible for the lensing effects). At the end of Subsection 3.4.1 we also pointed out that, because of the definition of the reference mass density Σ_c, given by Eq. (3.1), gravitational lensing effects are more easily observed when the lens is at a cosmological distance (i.e., at redshift typically > 0.1). This makes gravitational lensing a useful and complementary source of information on dark matter for distant systems for which traditional diagnostics (such as optical and radio spectroscopy or X-rays) become more difficult to apply (typically, because of lack of sensitivity and spatial resolution). Here we briefly summarize a few points of interest.

10.6.1 Strong Lensing

A very successful line of work[28] has focused on investigations in which stellar dynamical measurements are combined with strong lensing analyses to infer the density distribution in distant elliptical galaxies (out to redshifts $z \approx 1$). Samples of distant elliptical galaxies are selected based on the fact that they act as strong gravitational lenses, given the observation of arcs and multiple images associated with more distant sources along the same line of sight. The study of the gravitational lens configuration is accompanied by a photometric and spectroscopic study of the lens (to some extent analogous to the measurements of luminosity profiles and rotation curves in spiral galaxies; see Section 10.1) that makes it possible to decompose the contribution of dark and visible matter in the elliptical galaxy acting as lens (much like we described for the decomposition of the rotation curves in spiral galaxies, but here with the extra information

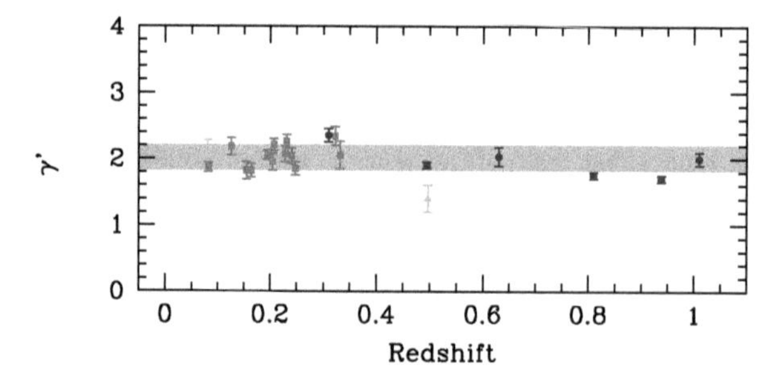

Figure 10.3 Logarithmic total mass (visible + dark) density slopes of field early-type galaxies as a function of redshift, derived from the combined use of stellar dynamics and gravitational lensing; the case $\gamma' = 2$ corresponds to the slope characteristic of the singular isothermal sphere (from: Koopmans, L. V. E., Treu, T., Bolton, A. S., Burles, S., and Moustakas, L. A., "The Sloan Lens ACS Survey: III. The structure and formation of early-type galaxies and their evolution since $z \sim 1$," 2006. *Astrophys. J.*, **649**, 599; reproduced by permission of the AAS).

provided by the observed lens effects). The very interesting result is that, if the total (stars + dark matter) density distribution of the lensing elliptical galaxy is modeled as a power law $\rho \sim r^{-\gamma}$, for relatively large samples of galaxies, visible and dark matter cooperate so that the relevant index turns out to be $\gamma \approx 2$. This is illustrated in Fig. 10.3.

As we have emphasized on many occasions, the r^{-2} behavior of the density profile is the signature of the self-gravitating isothermal sphere, which is associated with the gravitational field characteristic of flat rotation curves. In other words, with the help of gravitational lensing, it has been established that elliptical galaxies that are billions of light-years away share the same property of flat circular velocity profiles exhibited by nearby spiral galaxies (such as NGC 2403 and NGC 3198; see Fig. 9.3) and nearby bright elliptical galaxies (such as NGC 4472; see Fig. 10.1).

10.6.2 Weak Lensing

In clusters of galaxies, arcs and other strong lensing phenomena confirm that dark matter is more concentrated than visible matter (largely dominated by the IntraCluster Medium; see Subsection 8.1.3). This result is at variance with the case of dark matter halos in galaxies, which supports the picture in which (on the scale of individual galaxies) visible matter is more concentrated and dark matter has a more diffuse distribution. In addition, weak lensing analyses confirm that clusters of galaxies contain large amounts of dark matter, consistent with what was suggested earlier by X-ray studies.

Finally, we should at least mention that a number of projects are under way that aim to study the very weak lensing signal associated with the so-called cosmic shear.[29] To some extent, the information gathered by these projects should reflect that of the initial inhomogeneities (anisotropies) detected in the Cosmic Microwave Background radiation (see Section 3.2), thus exhibiting the evolution of structure formation from $z \approx 1000$ to $z \approx 1$.

10.7 An Alternative to Dark Matter

The concern expressed by Feynman (in a sentence that we quoted in Chapter 6 and recalled at the beginning of this chapter) is quite natural. Even though the law of gravitation has had wonderful confirmations on the scale of the solar system, we can legitimately ask whether for its applicability we have independent tests on the grand scales of galaxies and beyond. This point had already been raised long ago by eminent scientists.[30] When astronomers began to measure with sufficient accuracy the kinematics of galaxies, the issue was brought up again.[31]

Thus it is not surprising that at the time when evidence for the existence of dark halos was becoming established, some scientists explored the possibility that dark matter does not exist and the discrepancies between models and data gathered as proof of the existence of dark matter might be taken to indicate the need for a modification of the law of gravitation when applied to galactic scales. One well-thought-through attempt in this direction is called Modified Newtonian Dynamics (MOND). It was initially formulated in the 1980s and later was developed and improved in terms of empirical evidence and theoretical justification.[32] The purpose of this last section is to give an idea of the strategy behind this attempt and to point out some of its merits and limitations, with the general expectation that the attempt will eventually be abandoned once we identify what dark matter is made of. For this goal, we will not address some formal aspects and the theoretical foundation of MOND, but rather we will highlight some of the most intuitive features at the basis of this point of view.

For simplicity, let us go back to the asymptotic limit of Eq. (9.2), $V^2/r = GM(r)/r^2$, discussed in Chapter 9. We might argue that we can reconcile a flat rotation curve (a constant V) with the possibility that sufficiently far from the galaxy center the mass appearing in the equation $M(r)$ is a constant (the total mass of the galaxy, with no dark matter) if, at large distances, the law of gravitation changes from r^{-2} to r^{-1}. Of course, this "new theory" would require the introduction of a scale length r_{mo} (presumably on the order of 1 kpc, because we are focusing on the dynamics of galaxies), so that the above statements could be quantified as asymptotic behavior of the gravitational acceleration g in different regimes: $g \sim GM/r^2$ for $r \ll r_{mo}$ and $g \sim GM/(r_{mo}r)$ for $r \gg r_{mo}$. Note that a simple analytical expression such as $g = GM[(r + r_{mo})/(r_{mo}r^2)]$ would have the desired asymptotic limits; obviously, other functions could be considered. Based on this idea, we might decide to develop a suitable theoretical framework to support such a modified law of gravitation. In practice, this approach was discarded immediately as inconsistent with the observations. The main fact against this picture is that there are small galaxies (such as UGC 2259) that exhibit a significant mass discrepancy (i.e., the need for a dark halo in the description outlined in Chapter 9) and large galaxies (such as UGC 2885, with optical radius around 80 kpc) for which such mass discrepancy is not evident.

Another possibility is to consider the introduction of a scale acceleration a_0. Then, in order for the modified law of gravitation to meet the requirements set by the observed flat rotation curves, the following argument was proposed. On the right-hand side of Eq. (9.2), we replace the standard Newtonian gravitational acceleration g_N with a modified gravitational acceleration g, following the prescription $\mu(g/a_0)g = g_N$. Here parentheses indicate that the argument of the function μ is g/a_0 and the function $\mu(x)$ is chosen so as to satisfy the two asymptotic limits $\mu(x) \sim 1$ for $x \gg 1$ and $\mu(x) \sim x$ for $x \ll 1$. A simple function that obeys the two asymptotic limits is $\mu(x) = x(1 + x^2)^{-1/2}$. Thus in the regime in which the acceleration g is large (with respect to the scale a_0), we have that the right-hand side of Eq. (9.2) becomes $g \sim g_N$, that is, we are in the domain of Newtonian gravity. In turn, in the regime of small accelerations, $g \ll a_0$, the right-hand side of Eq. (9.2) becomes $g \sim \sqrt{a_0 g_N}$. With this choice, if the outer parts of galaxies turn out to fall in the regime of small accelerations, Eq. (9.2) implies $V^4 \sim GMa_0$, with M the total mass of the galaxy (with no dark matter). This possibility has been found to have no immediate counterindications from observations and is the one at the basis of MOND.

It has been shown that all high-quality rotation curves available at the end of last century[33] could be well fitted by MOND, based on a single parameter a_0 of the order of 10^{-8} cm s^{-2}. The value of a_0 obtained as a best-fit parameter changes slightly from galaxy to galaxy, but it has been argued that a single value could be acceptable if we allow for small changes of the adopted distances of the galaxies under investigation. Curiously, it happens that numerically $a_0 \approx H_0 c$, with H_0 the Hubble constant and c the speed of light. Figure 10.4 illustrates the quality of the fit for two well-known galaxies, NGC 2903 and NGC 3198.

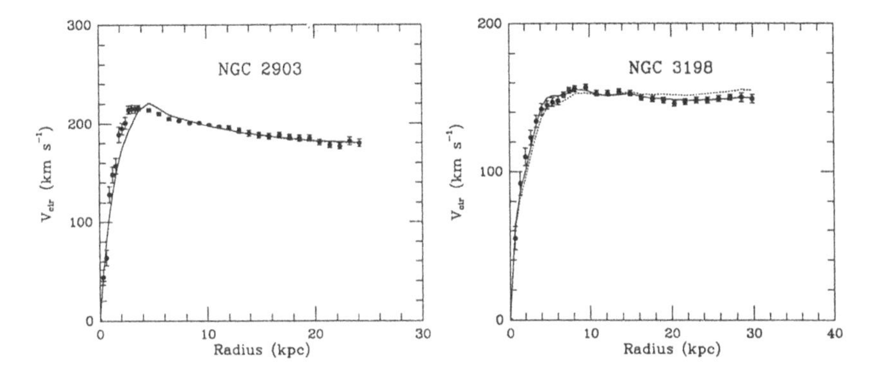

Figure 10.4 Two cases from a study of a sample of spiral galaxies (from: Begeman, K. G., Broeils, A. H., and Sanders, R. H., "Extended rotation curves of spiral galaxies: Dark haloes and modified dynamics," 1991. *Mon. Not. Roy. Astron. Soc.*, **249**, 523) in which the observed rotation curve is fitted without dark matter by considering a modified Newtonian dynamics. Two fits have been performed, a one-parameter fit (with *M/L* as free parameter) and a two-parameter fit for which the distance to the galaxy can also be adjusted slightly. The slightly discrepant fit for NGC 3198 is the one-parameter fit. The value of the threshold acceleration is taken to be fixed at $a_0 = 1.21 \times 10^{-8}$ cm s^{-2}.

A surprising aspect of the MOND approach is that it provides a natural explanation for the Tully–Fisher relation, which follows from $V^4 \sim GMa_0 = G(M/L)a_0L$, if we consider that for spiral galaxies the M/L ratio for the visible stellar disk is likely not to change significantly from galaxy to galaxy. (Of course, the first attempt that was initially discarded, based on the existence of a scale length r_{mo}, would not work in this respect, because it would predict the wrong slope for the Tully–Fisher relation, from $V^2 \sim GM/r_{mo}$.)

Another curious and intriguing point that has attracted the attention of scientists for more than a decade is a detailed study of the orbits of the two interplanetary probes *Pioneer* 10 and *Pioneer* 11. Because we have been in contact with the probes for more than 10 years, the orbits have been tracked with extremely high precision. Surprisingly, after a thorough analysis the orbits turned out to exhibit a small discrepancy with respect to the predicted orbits; it was shown that the discrepancy could be resolved by assuming the existence of an anomalous acceleration in the direction of the Sun on the order of a_0.[34] A reinvestigation of the entire set of data later suggested that the discrepancy might be due to some subtle thermal effects on the spacecraft that had been previously overlooked, with no need for invoking the presence of anomalous accelerations.[35]

MOND has been challenged and tested to confront a large variety of empirical and theoretical issues. In this introductory book, it would be inappropriate to give a detailed account of all these arguments. In closing, we only mention one astronomical discovery that probably provides the strongest empirical evidence against MOND. This is the study of the Bullet cluster of galaxies (1E 0657-56; the cluster is at redshift $z = 0.296$).[36] The structure of the cluster (see Fig. 4.2) is highly complex and is thought to result from the ongoing merger of two separate clusters of galaxies. It has been studied in great detail with optical and X-ray telescopes. The optical images show the distribution of galaxies in the sky, whereas the X-ray images show the corresponding distribution of the IntraCluster Medium. We recall that the ICM is thought to be the dominant component of the visible mass in clusters of galaxies (see Section 8.1). Finally, a weak-lensing analysis on the distortions of thousands of more distant galaxies along the same line of sight produced by the Bullet cluster as a gravitational lens has been carried out and has led to the measurement of the distribution of the total mass (visible plus dark) in the sky. The key point is that, in the sky, the center of the visible mass (dominated by the ICM) is offset from the center of the total mass (dominated by dark matter).[37] The offset appears to rule out any explanation in terms of a modification of the law of gravitation and instead is strong evidence for the very existence of dark matter. Other cases, such as the Ring cluster[38] and the cluster A520 (at redshift $z = 0.201$), appear to set additional examples in this context.

Notes

1 For example, see Schechter, P. L. 1990. In *Dynamics and Interactions of Galaxies*, ed. R. Wielen, Springer-Verlag, Berlin, p. 508.
2 The interested reader may consult Ciotti, L. 2021. *Introduction to Stellar Dynamics*. Cambridge University Press, Cambridge, UK; Bertin, G. 2014. *Dynamics of Galaxies*, 2nd ed. Cambridge University Press, New York, NY, and the many references provided there.
3 However, deep 21-cm observations of early-type galaxies have identified the presence of low density HI disks, sometimes regular and radially extended; see Serra, P., Oosterloo, T., et al. 2012. *Mon. Not. Roy. Astron. Soc.*, **422**, 1835.
4 Davies, R. L., Birkinshaw, M. 1988. *Astrophys. J. Suppl.*, **68**, 409.
5 Tonry, J. L. 1983. *Astrophys. J.*, **266**, 58.
6 Lynden-Bell, D. 1967. *Mon. Not. Roy. Astron. Soc.*, **136**, 101; van Albada, T. S. 1982. *Mon. Not. Roy. Astron. Soc.*, **201**, 939.
7 Bertin, G., Bertola, F., et al. 1994. *Astron. Astrophys.*, **292**, 381; Gerhard, O., Kronawitter, A., et al. 2001. *Astron. J.*, **121**, 1936.
8 For NGC 3379, compare the results of Fukazawa, Y., Botoya-Nonesa, J. G., et al. 2006. *Astrophys. J.*, **636**, 698 with those of Trinchieri, G., Pellegrini, S., et al. 2008. *Astrophys. J.*, **688**, 1000.

9 Romanowsky, A. J., Douglas, N. G., et al. 2003. *Science*, **301**, 1696.
10 For a thorough introduction to cosmology written before the discovery of the acceleration of the universe, see Peebles, P. J. E. 1993. *Principles of Physical Cosmology*. Princeton University Press, Princeton, NJ.
11 To answer this question, some projects have been carried out making use of gravitational lensing, in the form called galaxy–galaxy lensing. See Griffiths, R. E., Casertano, S., et al. 1996. *Mon. Not. Roy. Astron. Soc.*, **282**, 1159; for a more recent reference, see Parker, L. C., Hoekstra, H., et al. 2007. *Astrophys. J.*, **669**, 21.
12 Bahcall, N. A. 1988. *Annu. Rev. Astron. Astrophys.*, **26**, 631. See also the interesting article by Bahcall, N. A., Fan, X. 1998. *Astrophys. J.*, **504**, 1 and the review by Rosati, P., Borgani, S., Norman, C. 2002. *Annu. Rev. Astron. Astrophys.*, **40**, 539.
13 Springel, V., White, S. D. M., et al. 2005. *Nature (London)*, **435**, 629.
14 Springel, V., Yoshida, N., White, S. D. M. 2001. *New Astronomy*, **6**, 79.
15 Xu, G. A. 1995. *Astrophys. J. Suppl.*, **98**, 355; Barnes, J., Hut, P. A. 1986. *Nature (London)*, **324**, 446.
16 Vogelsberger, M., Genel, S., et al. 2014. *Mon. Not. Roy. Astron. Soc.*, **444**, 1518.
17 *IllustrisThe Next Generation*; Pillepich, A., Springel, V., et al. 2018. *Mon. Not. Roy. Astron. Soc.*, **473**, 4077.
18 Navarro, J. F., Frenk, C. S., White, S. D. M. 1996. *Astrophys. J.*, **462**, 563. In this article, physical quantities are all referred to a choice of Hubble's constant $h = 0.5$.
19 The subscript E is used to recall the name of Jaan Einasto; for example, see Einasto, J. 1965. *Trudy Inst. Astrofiz. Alma-Ata*, **5**, 87; Einasto, J. 1969. *Astron. Nachr.*, **291**, 97; Einasto, J., Haud, U. 1989. *Astron. Astrophys.*, **223**, 89; Merritt, D., Graham, A., et al. 2006. *Astron. J.*, **132**, 2685; Chemin, L., de Blok, W. J. G., Mamon, G. A. 2011. *Astron. J.*, **142**, 109. The title of Einasto's 1969 article "On galactic descriptive functions" and the abstract of his 1989 article with Haud clearly emphasize the descriptive character of the proposed models.
20 As we noted in Section 10.1, an answer to the problem of the universality of the $R^{1/4}$ law of bright elliptical galaxies, interpreted as the signature of incomplete violent relaxation, has been proposed; for further details, see Chapter 22 of Bertin, G. 2014. op. cit. and references therein.
21 Kahn, F. D., Woltjer, L. 1959. *Astrophys. J.*, **130**, 705.
22 van der Marel, R. P., Fardal, M., et al. 2012. *Astrophys. J.*, **753**, id.8.
23 We recall that after his initial suggestion (Lynden-Bell, D. 1976. *Mon. Not. Roy. Astron. Soc.*, **174**, 695), Lynden-Bell argued that certain streams of globular clusters would trace the orbits of satellites that may have merged with the Galaxy and predicted the proper motions for 22 objects as a test of this scenario; Lynden-Bell D., Lynden-Bell, R. M., 1995. *Mon. Not. Roy. Astron. Soc.*, **275**, 429.
24 Pawlowski, M. S., Kroupa, P. 2020. *Mon. Not. Roy. Astron. Soc.*, **491**, 3042.
25 The Large Magellanic Cloud, the Small Magellanic Cloud, Draco, Ursa Minor, Fornax, Leo II, and the Carina dwarf spheroidal; Sculptor orbits along the same plane, but in the opposite direction.
26 van den Bergh, S. 2008. *Mon. Not. Roy. Astron. Soc.*, **390**, L51.
27 Lokas, E. L., Mamon, G. A., Prada, F., 2005. *Mon. Not. Roy. Astron. Soc.*, **363**, 918; Strigari, L. E., Bullock, J. S. et al. 2008. *Nature (London)*, **454**, 1096; Strigari, L. E., Koushiappas, S. M., et al. 2008. *Astrophys. J.*, **678**, 614; Adams, J. J., Simon, J. D., et al. 2014. *Astrophys. J.*, **780**, id.63.
28 Starting with Koopmans, L. V. E., Treu, T. 2002. *Astrophys. J. Lett.*, **568**, L5; Treu, T., Koopmans, L. V. E. 2002. *Astrophys. J.*, **575**, 87; followed by many papers.
29 For example, see Simpson, F., Heymans, C., et al. 2013. *Mon. Not. Roy. Astron. Soc.*, **429**, 2249.
30 See Jeans, J. H. 1923. *Mon. Not. Roy. Astron. Soc.*, **84**, 60.

31 Finzi, A. 1963. *Mon. Not. Roy. Astron. Soc.*, **127**, 21; the author refers explicitly to the 1957 measurement of the rotation curve of M31 by van de Hulst, Raimond, and van Woerden that we mentioned in Chapter 9 and showed in Fig. 3.1.
32 Milgrom, M. 1983. *Astrophys. J.*, **270**, 365; Bekenstein, J. D., Milgrom, M. 1984. *Astrophys. J.*, **286**, 7. See also Sanders, R. H. 2010. *The Dark Matter Problem: A Historical Perspective*. Cambridge University Press, Cambridge, UK. The theory has been given a formal framework as an alternative theory of gravitation; Bekenstein, J. D. 2004. *Phys. Rev. D*, **70**, id.083509.
33 Begeman, K. G., Broeils, A. H., Sanders, R. H. 1991. *Mon. Not. Roy. Astron. Soc.*, **249**, 523. See also Sanders, R. H., Verheijen, M. A. W. 1998. *Astrophys. J.*, **503**, 97.
34 Anderson, J. D., Laing, P. A., et al. 2002. *Phys. Rev. D*, **65**, id.082004.
35 Anderson, J. D., Morris, J. R., et al. 2012. *Phys. Rev. D*, **85**, id.084017.
36 Clowe, D., Bradač, M., et al. 2006. *Astrophys. J. Lett.*, **648**, L109.
37 How this could happen has been explained in terms of the properties of the dynamics of the merger of two clusters; but the correctness of this explanation is of secondary importance in the argument against MOND.
38 Cl 0024+17, at redshift $z \approx 0.4$; see Jee, M. J., Ford, H. C., et al. 2007. *Astrophys. J.*, **661**, 728.

Index

Printed in the USA
CPSIA information can be obtained
at www.ICGtesting.com
LVHW020506090923
757629LV00005B/102